Evolving Households

Evolving Households

The Imprint of Technology on Life

Jeremy Greenwood

The MIT Press
Cambridge, Massachusetts
London, England

This book was set in Stone Serif by Westchester Publishing Services. Printed and bound in the United States of America.

Library of Congress Cataloging-in-Publication Data
Names: Greenwood, Jeremy, author.
Title: Evolving households : the imprint of technology on life / Jeremy Greenwood.
Description: Cambridge, MA : MIT Press, [2018] | Includes bibliographical references and index.
Identifiers: LCCN 2018016802 | ISBN 9780262039239 (hardcover : alk. paper)
Subjects: LCSH: Technological innovations—Economic aspects—United States—History. | Families—United States—History. | United States—Social conditions.
Classification: LCC HC110.T4 G694 2018 | DDC 303.48/30973—dc23
LC record available at https://lccn.loc.gov/2018016802

10 9 8 7 6 5 4 3 2 1

To Amy, who knows more about the family than I ever will.

Contents

Acknowledgments   ix

Introduction   xi

1   More Working Mothers   1
    1   Measuring the Allocation of Time   2
    2   The Unisex Single Model of Labor Supply   5
    3   Married Female Labor Supply   25
    4   Calibrating Labor Supply Models   35
    5   Household Production Theory   37
    6   The Rise in Married Female Labor Supply   45
    7   The Gender Wage Gap   62
    8   Brain and Brawn   65
    9   The Division of Labor in the Household   69
    10   Literature Review   72
    11   Problems   76

2   The Baby Boom and Baby Bust   79
    1   Definitions of Fertility   80
    2   Fertility, 1800–1990   83
    3   The Mystery of the Baby Boom   86
    4   A Basic Model of Fertility   89
    5   The Size and Start of the Baby Boom in OECD Countries   100
    6   Advances in Obstetric and Pediatric Medicine   103
    7   Fertility and Wars: The Case of World War I in France   108
    8   The Choice between Jobs and Kids   112
    9   Malthus   117
    10   Literature Review   122
    11   Problems   127

3   The Decline in Marriage   129
    1   Love or Money   130
    2   Definitions for Marriage and Divorce Rates   131

　3　Trends in Marriage and Divorce　132
　4　A Basic Model of Marriage　140
　5　Divorce　151
　6　Assortative Mating　156
　7　Growing Up with a Single Mother　162
　8　The Beckerian Theory of Marriage　170
　9　Literature Review　178
10　Problems　179

4　Social Change　181
　1　Women's Rights in the Workplace　184
　2　Mothers and Sons　191
　3　The Sexual Revolution, 1900–2000　195
　4　Measures of Nonmarital Births　196
　5　Technological Progress in Contraception　198
　6　A Model of Premarital Sex　204
　7　The Socialization of Children　208
　8　The Frequency of Sex, a Digression　220
　9　The Spirit of Capitalism　223
10　Literature Review　230
11　Problems　232

5　Increased Longevity and Longer Retirement　235
　1　Better Health　236
　2　The Development of New Drugs　243
　3　Health Insurance　247
　4　AIDS　251
　5　The Trend in Retirement　255
　6　Old-Age Social Security　263
　7　Literature Review　266
　8　Problems　267

6　Conclusion　269

Mathematical Appendix　275
1.A　Maximizing a Function　275
2.A　Total Differentials　281
3.A　Leibniz's Rule　282
4.A　Distribution Functions and Correlation Coefficients　282
5.A　A Linear First-Order Difference Equation　286

References　291
Index　301

Acknowledgments

Many generations of students at the University of Pennsylvania have been subjected in the classroom to versions of the book. Feedback from the students came in many forms: questions in class, pointing out errors and shortcomings, and sometimes just the looks on faces. As a result the presentation has evolved over time. Students at the Universität Konstanz also provided some comments during a short course taught there. I am especially grateful to my former graduate students at the Universities of Pennsylvania and Rochester who followed me into the area of family economics. They never questioned the wisdom of the pursuit, which at times must have seemed a risky venture. All have published their own research in this area, as well as collaborated on research with me. The older generations of them are now distinguished researchers, while the younger ones are off to a good start. So, thanks go to Nezih Guner, Karen A. Kopecky, Cezar Santos, Ananth Seshadri, Guillaume Vandenbroucke, David Weiss, and Mehmet Yorukoglu. I am proud of all of them. Nezih Guner, in particular, has accompanied me on the adventure from the very start and has continued working with me until this day, so no amount of thanks can repay the debt I owe to him.

Emily Taber, the acquisitions editor at MIT Press, read the entire manuscript and provided many great comments, both macro and micro, that have been incorporated into the book. Several reviewers had useful suggestions as well, which substantially changed the format of the book. Last, I am grateful to Thomas J. Sargent, who introduced me to MIT Press.

Introduction

When one thinks about technological progress, the First Industrial Revolution may come to mind. One might imagine a steam engine in a factory running many whirling machines via a system of belts and wheels. For the Second Industrial Revolution it might be dams with rushing water powering dynamos supplying electricity to industry, goods being trucked along highways to stores, and the sky filled with aircraft. It is also easy to visualize microchips packed with transistors processing torrents of zeros and ones, perchance running robots, as typify the Information Age. The impact of technological progress on industry has been great. Its effect on the household and culture is just as significant, though. This is the subject here. The book develops economic models to study the rise in married female labor-force participation, the drop in fertility and the baby boom, the decline in marriage, women's liberation, the increase in premarital sex, the upswing in life expectancy and expenditure on health care, and the movement toward spending more time in retirement. All of these trends characterize a dramatic transformation of everyday life made possible by advancements in technology.

In the 1800s the mother in most American households worked at home surrounded by six children. Housework was laborious in a world without running water, central heating, and electricity. The Second Industrial Revolution introduced electricity and labor-saving household appliances. Additionally, the value of physical strength declined on the labor market as machinery took over strenuous tasks. It is not a stretch to say that these developments liberated married women from the home. The rise in married female labor-force participation, as a function of such technological progress, is addressed in chapter 1. The chapter also touches upon the division of labor in a married household, both empirically and theoretically, and how this has changed over time.

In little over 150 years the number of siblings a person had dropped from six to two. This is the subject of chapter 2. Technological advance implies that a unit of time becomes more valuable in terms of the goods that it can produce. As the value of time rose, so did the opportunity cost of having children. This led to a secular decline in fertility. The drop was punctuated by the baby boom, however. It is argued that the revolution in labor-saving household technologies, in conjunction with advances in obstetrics and pediatrics, reduced the cost of having children around World War II and led to a temporary increase in fertility.

Almost no unmarried young women (18–30 years old) lived alone independently in 1900. Today about half do. There has been a dramatic fall in the fraction of the population that is married and a concurrent rise in the proportion that have been divorced. Technological progress in the home and market sectors has meant that people can afford to be more choosey about whom they marry; this was a luxury that earlier generations could not afford. Now, relative to the past, economic necessity matters less as a reason for marriage, and romantic considerations count for more. Chapter 3 discusses these developments.

Married women were prohibited from taking many jobs, such as teaching, in the first part of the twentieth century. As technological progress in the home and the marketplace changed the value of a woman's labor at home relative to the market, an increased demand for women's rights occurred. This is perfectly exemplified by Elizabeth Cady Stanton's prescient 1892 testimony to the U.S. Congressional Committee on the Judiciary, in which she noted that machinery was lifting the burden of housework off of the shoulders of women (she is quoted in chapter 1). Chapter 4 analyzes how culture and social norms are affected by technological advance. Culture and social norms in turn influence individuals' behavior and hence the economy. In other words, the economy affects culture and culture the economy. The chapter starts off with a voting model illustrating how women are granted rights in the workplace as economic circumstances change. Technological progress manifests itself in many forms. One is innovation in contraception. This had a profound impact on sexual mores, as is discussed. Parents and social institutions try to inculcate many norms in children, such as not to engage in risky behaviors or to be patient and work hard. One way they may do this is by molding children's tastes. The chapter illustrates how the means and ends may change as the economic environment evolves.

In 1850 a newly born American could expect to live half as long as one born today. Furthermore, at that time, the vast majority of men aged 75–79 still worked. Advances in medicine have extended life. This is the subject of chapter 5. Here a model of investment in health is developed. It is shown how technological advance in the health-care sector leads to prolonged life and increased expenditure on health care as a fraction of income. The chapter also discusses the development of new drugs by firms. As life spans lengthened, the fraction of life spent in retirement expanded. The chapter shows how people will spend a larger fraction of their lives in retirement, both as standards of living improve and the price of the goods consumed in retirement drop.

Who would have thought that economics can shed light on such topics? The book treats this subject matter from a macroeconomic perspective. It focuses on addressing long-run trends, as does growth theory. Here, too, the engine of change is technological progress. The vehicles of analysis are versions of the macroeconomist's representative household model, starting at the level of specifying tastes and technology.

The book has been written in such a way as to make the ideas it contains accessible to everyone from advanced undergraduates to graduate students and researchers. It requires knowledge of elementary calculus and a very limited amount of probability theory. The mathematics needed are provided in the Mathematical Appendix, so the book should be self-contained at the advanced undergraduate level. For an advanced undergraduate the payoff from this little investment in mathematics is huge. Surely, it is time for her or him to make this small investment and reap the large returns from the world of modern economics. For graduate students the book is chock-full of ideas. In an age when the vast majority of graduate students and researchers are literate in mathematics, an idea is everything. Often the successful transliteration of a idea into an economic model is very simple. Economic models delineate in a precise fashion the relationships between variables that are taken to be exogenous—say, technological progress—and ones that are assumed to be endogenous—maybe married female labor-force participation. As an economic model gets refined and generalized, perhaps to match features of the data, it grows more complicated in its formulation. The book illustrates how the initial incarnation of an idea in an economic model is often not very complicated. It's a valuable skill to be able to distill an idea down to its core, especially when starting a research project. This book illustrates how this can be done.

Gender issues are at the core of many of the ideas presented in this book. These are addressed in a disinterested scientific manner. Economic models are abstractions and, as such, cannot incorporate all features of the real world. Historically speaking, married women did not work outside of the home. They still work less in the market and more at home than men do, as the evidence on American's uses of time presented in chapter 1 shows. Because the book focuses on long-run trends, often the vehicle of analysis is a traditional married household, in which the man always works full time, while the woman's time is split between the home and the workplace. Of course, these days most women work in the workplace and men do a little more housework than in the past. Also, couples may not be married, and they may be of the same sex. Facts on these newer forms of relationships are not abundant. So, confronting theories about them with evidence is harder. Still, much of the theoretical analysis is germane for these newer forms of households, as is made clear.

Additionally, women earn less in the labor market than men do. Discrimination is an obvious explanation. The difference in pay may also have something to do with the fact that women choose different jobs than men because their careers may be interrupted to raise children. In parts of the book the gender wage gap is taken as exogenous for simplicity, while in other parts it is modeled endogenously, as a function of discrimination, the choice between kids and jobs, and the comparative advantage that women have in certain jobs as a function of technological development. A child in a single-parent family is much more likely to be raised by her or his mother than father, as the analysis presumes. Finally, there was, and still is, a double standard concerning premarital sex. This may have arisen because the cost of an unwanted pregnancy is higher for a woman than for a man; even today, many men don't pay child support or devote much time to their out-of-wedlock children. So, it is probably fair to say that advances in contraception have affected women more than men. Hence, to keep things simple, the analysis of the decision to become sexually active is framed from a young woman's perspective. Of course, times are changing, and so is the position of men and women in society. But the changing nature of everyday life is precisely the subject of this book.

1   More Working Mothers

The central goal of this introductory chapter is to analyze the dramatic increase in married female labor-force participation over the course of the twentieth century. The hypothesis is that technological advance in the home and in the market caused this increase. To set the stage, the chapter starts with some facts on the amount of time that Americans, both men and women, have spent on market work, nonmarket work, and leisure during the postwar period. It then turns to developing models of labor supply for understanding such facts.

Toward this end, the formal analysis begins with the standard unisex single model of labor supply, which will be familiar to many. Once this benchmark is established, the household labor supply decision for a married woman in a traditional two-person family is addressed. Labor supply for a married women is modeled, in turn, along both the intensive and extensive margins, concepts that will be defined later. Household production theory is then introduced. This theory treats the household as a small factory that uses intermediate goods and labor to produce nonmarket goods. The decision of a household to adopt labor-saving household inputs for their factory, as a function of the input price, is formulated.

Attention is then directed to the main topic of the chapter: namely, the rise in married female labor-force participation. The analysis focuses on three potential causes: technological progress in the household sector due to the introduction of labor-saving household appliances and subsequent drops in their prices; industrialization; and a narrowing of the gender wage gap, or an increase in the ratio of women's to men's wages. Historical evidence about the rise in married female labor-force participation, the decline in housework, and the spread of modern appliances is presented. The labor supply model with household production is then used to address the dramatic rise in labor-force participation by married women over the course of the twentieth century.

After doing this, some evidence on the gender wage gap is presented. Then, a model of industrialization is developed to show how the gender wage gap narrows over time as a result of the economy transforming from a primitive one, in which physical strength is valued, to a more advanced one, where mental skills are prized. Such a shift favors women relative to men. The penultimate section addresses the division of labor between a married couple. The chapter concludes with a brief review of the relevant literature.

1 Measuring the Allocation of Time

Books such as Juliet Schor's *The Overworked American: The Unexpected Decline of Leisure* or Ruth Schwartz Cohen's *More Work for Mother: The Ironies of Household Technology from the Open Hearth to the Microwave* suggest that Americans, especially women, have never worked so hard and enjoyed so little leisure. These books are long on words and short on facts. So, do people work more today than in the past? Do women work harder than men? For many years now, social scientists have surveyed Americans' uses of time. These surveys can be used to answer such questions. In an effort to do so, Aguiar and Hurst (2007) patch the various surveys together to obtain a picture of what has happened to time allocations over the last five decades. The first survey, done in 1965–1966, focused on 2,001 individuals, while the last one, performed in 2003, questioned 20,720 people. Aguiar and Hurst (2007) conclude:

(1) Hours worked for men in the labor market has decreased significantly over the period 1965 and 2003.

(2) Hours worked for women in the labor market has increased over the same time frame.

(3) Hours worked at home by women has decreased significantly.

(4) Leisure has increased significantly for both men and women.

(5) Men and women enjoy about the same amount of leisure.

Table 1.1 shows figures for core market work, which is defined by Aguiar and Hurst (2007) to be "all time spent working in the market sector on main jobs, second jobs, and overtime, including any time spent working at

Table 1.1
Hours per week spent working by men and women.

	1965	1975	1985	1993	2003	Difference 2003–1964	
Men							
Core market work	42.09	39.80	36.86	38.52	35.54	−6.55	−16.9%
Total market work, m	51.58	46.53	43.35	42.74	39.53	−12.05	−26.6%
Core nonmarket work	1.96	2.01	3.82	2.90	3.40	1.44	55.1%
Total nonmarket work, n	9.67	10.85	13.96	12.44	13.43	3.75	32.9%
Child care, c	1.44	1.40	1.66	1.47	3.24	1.80	22.3%
Total work, m+n+c	62.69	58.78	58.97	56.65	56.20	−6.49	−10.9%
Sample size	833	756	1,412	2,483	6,699		
Women							
Core market work	18.33	19.24	19.84	22.49	22.65	3.82	21.2%
Total market work, m	22.45	22.74	23.41	24.97	24.93	2.48	10.5%
Core nonmarket work	22.61	19.43	16.89	13.83	13.23	−9.38	−53.6%
Total nonmarket work, n	32.86	28.21	27.10	23.56	22.55	−10.31	−37.7%
Child care, c	5.60	4.60	5.36	4.54	7.46	1.86	28.7%
Total work, m+n+c	60.91	55.55	55.87	53.06	54.94	−5.97	−10.3%
Sample size	1,021	917	1,756	2,864	8,392		

See the main text for the definitions of the time-use variables.
Source: Aguiar and Hurst 2007, table 2.

home." For men, it fell from 42 hours per week to 35.5, a drop of 17 percent (measured in terms of the difference between the natural logarithms of the data points). For women it increased from 18.83 to 22.65 hours per week, a rise of 21 percent. Total market work adds in commuting and break times. Core nonmarket work is defined by Aguiar and Hurst (2007) as "any time spent on meal preparation and cleanup, doing laundry, ironing, dusting, vacuuming, indoor household cleaning, and indoor design and maintenance (including painting and decorating)." Total nonmarket work adds in the time spent obtaining goods and services, home maintenance, outdoor cleaning, vehicle repair, gardening, and pet care. Core nonmarket work for men rose from 1.96 to 3.40 hours, an upward movement of 55 percent. For women, it dropped from 22.61 to 13.23 hours, or 54 percent. Still, in 2003 women did roughly 4 times as much core nonmarket work as men. Total time spent working adds together total market and nonmarket work, including time spent on child care. For men this moved down from

62.69 hours to 56.20, while for women it dropped from 60.91 to 54.94. This represents declines of 11 and 10 percent, respectively. Total time spent working is roughly the same for each sex.

The other side of the coin is leisure. As Becker (1965) notes, giving a precise definition for leisure is problematic. Some people enjoy cooking, gardening, and so on, while others hate it. An attempt is made to address this problem by presenting several measures for leisure. The narrowest measure "sums together all time spent on 'entertainment/social activities/relaxing' and 'active recreation' (Aguiar and Hurst 2007)." Included here are activities such as reading for enjoyment, relaxing, socializing with friends, watching television, and the like. By this measure, men enjoy more leisure than women. As shown on table 1.2, men saw their leisure rise by 16 percent, from 31.80 to 37.40 hours. Women realized a 12 percent increase in leisure, from 29.89 to 33.54 hours. When the notion of leisure is expanded (measure 2) to include "time spent sleeping, eating, and on personal care," men and women spend about the same time. Measure 3 includes time spent on child care, since parents rank time spent with their children, say playing or reading, as among the most enjoyable things they do. The broadest measure, 4, is just time not spent working, either in the market or

Table 1.2
Hours per week spent on leisure by men and women.

	1965	1975	1985	1993	2003	Difference 2003–1965	
Men							
Leisure measure 1	31.80	33.36	35.15	37.65	37.40	5.60	16.2%
Leisure measure 2	101.68	105.33	106.81	108.50	107.88	6.20	5.9%
Leisure measure 3	103.12	106.73	108.47	109.97	111.13	8.01	7.5%
Leisure measure 4	106.75	110.62	110.68	112.82	115.04	8.29	7.5%
Women							
Leisure measure 1	29.89	33.14	34.46	37.32	33.54	3.65	11.5%
Leisure measure 2	102.70	107.75	108.69	111.38	107.59	4.89	4.7%
Leisure measure 3	108.31	112.35	114.05	115.92	115.06	6.75	6.1%
Leisure measure 4	112.69	117.05	117.49	119.48	120.52	7.83	6.7%

See the main text for the definitions of the leisure variables.
Source: Aguiar and Hurst 2007, table 3.

at home. Women have even more leisure than men when the broader measures (3 and 4) are used. Note that all measures show a significant rise in leisure for both sexes.

Furthermore, Aguiar and Hurst (2007) report that the average time that both men and women spent on total market work dropped from 35.98 to 31.71 hours per week over the period 1964 to 2003, a decline of 12.6 percent. (This average decline occurred despite the fact that women increased their time working in the market.) One might think that some poorer countries today are in a situation similar to that of the United States in the past, say the first part of the twentieth century. This is true in the sense that some poorer countries have income levels similar to those of the United States of yesteryear. Still, these countries now have access to technologies that were not available in the past. So, it is not clear to what extent a poor country today might resemble the United States of the past. The cross-country relationship between average market hours worked and the natural logarithm of real GDP per capita is plotted in figure 1.1. As can be seen, average market hours and real per-capita GDP are negatively associated. The Pearson correlation coefficient (a measure of the linear association between two series, discussed in the Mathematical Appendix) is -0.64 between the two series. So, time spent working in the market declines with economic development. Thus, it appears that leisure is a luxury good. The chapter now turns to models of labor supply that can be used to address facts such as these.

2 The Unisex Single Model of Labor Supply

The standard model of consumption-leisure choice focuses on a single person and does not distinguish between men and women. It is a *unisex single model*. This model provides a useful benchmark to compare with models of a married household's consumption-leisure choice.

2.1 The Standard Consumption-Leisure Diagram
Figure 1.2 illustrates the unisex single model. The vertical axis gives the person's consumption, c. The horizontal axis shows leisure, l. Suppose that the person has one unit of time that can be devoted to labor,

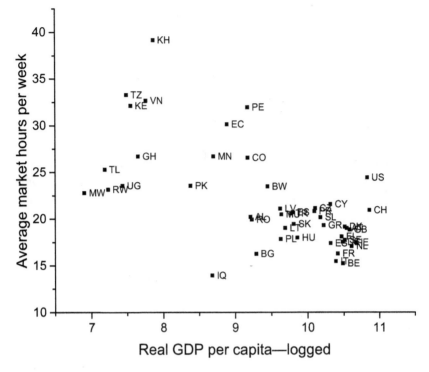

Figure 1.1
The cross-country relationship between average market hours per week and the logarithm of GDP per capita. A nation is labeled by its standard two letter country code.
Source: Bick, Fuchs-Schundeln, and Lagakos 2018.

$h = 1 - l$, or leisure, l. Each unit of time spent working is rewarded by the wage w.

2.1.1 The budget line   The line BB gives the person's budget constraint. It is described by the linear equation

$c + wl = w.$

The equation states that the amount the person spends on consumption, c, and leisure, wl, must equal potential income, w. The budget constraint can be rewritten as

$c = w(1 - l).$

Consumption, c, equals labor income, $w(1 - l)$, which is simply the product of the wage rate, w, and hours worked, $1 - l$. The budget line gives all the

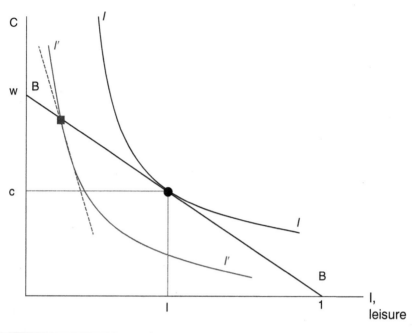

Figure 1.2

The consumption-leisure decision. At the optimal leisure-consumption point the marginal rate of substitution (MRS) of leisure for consumption is set equal to the wage rate so that $MRS = w$.

consumption-leisure combinations that the person can potentially enjoy with their one unit of time.[1] The slope of the budget constraint is given by (the negative of) the wage rate, w. Think about this as representing the relative price of leisure in terms of consumption: if the person desires to enjoy an extra unit of leisure, they must sacrifice w in terms of consumption. If the person worked all of their time, so that leisure was zero (or $l = 0$), then consumption would be w. This is the vertical intercept of the line. If the individual didn't work, then leisure would equal 1, implying $l = 1$. That person would of course have no consumption, so $c = 0$. This is the horizontal intercept.

1. They can also enjoy the points lying inside the line but would never choose to do so. That is, the budget constraint could be rewritten more generally as

$c + wl \leq w.$

The fact that they will pick only a leisure-consumption bundle, (l, c), lying on the budget line will become clear shortly.

2.1.2 Indifference curves The individual likes both consumption and leisure. Every leisure-consumption point, (l, c), in figure 1.2 is associated with a *unique* level of utility. The line II represents an indifference curve for an individual. It gives the combinations of consumption, c, and leisure, l, that generate some particular level of utility, u. Indifference curves have four properties:

Property 1 *Higher indifference curves have higher utility.* The farther out from the origin an indifference curve is, the higher is the level of utility, u, that is connected with it. This occurs because people enjoy both consumption and leisure; both are "goods." Move outward along a ray from the origin. Since more goods are better than fewer goods, utility must increase along any ray.

Property 2 *Indifference curves cannot cross one another.* At any point of intersection the two indifference curves must have the same level of utility, because every leisure-consumption point (l, c) is associated with a *unique* level of utility. This leads to a contradiction. To see this, note that for some points to the immediate right (or left) of an intersection point, one of the indifference curves will lie above the other. This implies that at these points the utility connected with the higher indifference curve must be larger for any common level of leisure because it offers more consumption. This contradicts the assumption that utility is constant along an indifference curve.

Property 3 *Indifference curves slope downward.* Why? Again, along an indifference curve utility is fixed at u. Therefore, to give the person more leisure, l, you must take away some of their consumption, c, at least if you want to keep them at the specified level of utility, u. The (negative of the) slope of the indifference curve gives the *marginal rate of substitution* of leisure for consumption. In other words, it specifies the amount of consumption that the person is willing to forgo in order to gain an extra unit of leisure, while maintaining their current level of utility. Any more consumption would reduce the person's utility and any less would raise it.

Property 4 *The slope of an indifference curve decreases* (in absolute value) as you move from left to right along the horizontal axis. The more leisure a person enjoys the less consumption they are willing to give up for yet an extra unit of leisure. This reflects *diminishing marginal utility* in leisure and consumption. Each incremental unit of leisure generates less and less in extra utility. Likewise, each marginal unit of consumption that is taken away results in increasing losses in utility.

2.1.3 The equilibrium consumption-leisure combination

An equilibrium for the person occurs at the point where the indifference curve, shown by II, is tangent to the budget line, portrayed by BB. At this point the slope of the indifference curve is equal to the slope of the budget line. The (absolute value of the) slope of the indifference curve is the marginal rate of substitution (MRS) of leisure for consumption, while (the absolute value of) the slope of the budget line is the wage rate, w. Suppose the marginal rate of substitution of leisure for consumption is higher than the wage rate, as indicated by the square. By working one unit less, the individual will give up w units of consumption, according to the budget constraint. Their marginal rate of substitution is bigger than this number, implying that the person is willing to give up more than w units of consumption to gain an extra unit of leisure. Hence, the individual should work less. By moving to the point indicated by the circle, they would increase utility, by Property 1. A point on a higher indifference curve is not feasible. It lies outside of the budget line. Therefore, it is too expensive for the person. A point on a lower indifference curve yields a lower level of utility, again by Property 1.

2.1.4 The impact of a change in wages

What happens when the wage rate, w, changes? Specifically, assume that it rises from w to w', as shown in figure 1.3. The rise in the wage rate causes the budget constraint to rotate outward from BB to $B'B$. The person's equilibrium moves to the new point (l', c'). In the situation portrayed, consumption rises while leisure falls. In general, however, the impact of a wage increase on leisure is *ambiguous*.

To see this, the above move is broken down into two steps, as shown in figure 1.4. Let the person labor at the new wage, w', but reduce either their endowment of time or income so that the new budget line passes through the old leisure-consumption point (l, c). In particular, you need to take away exactly $(w' - w)(1 - l)$ in income or $(w' - w)(1 - l)/w'$ in time. This leads to a parallel downward shift in the budget constraint from $B'B$ to $B''B''$ that goes through the point (l, c). This is called a *Slutsky compensated shift in income*. To prove that it goes through the original leisure-consumption point (l, c), note

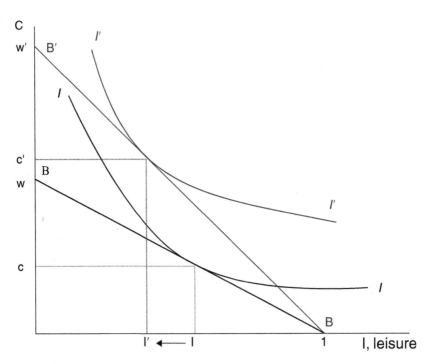

Figure 1.3
The impact of a rise in the real wage from w to w'.

that following the (negative) compensation, the person's budget constraint
is given by

$$c' + w'l' = w' - (w' - w)(1 - l),$$

where the term $-(w' - w)(1 - l)$ on the right-hand side is the Slutsky
compensation. Can the individual still buy their old leisure-consumption
bundle following the Slutsky compensation? It will now be shown that the
old leisure-consumption point (l, c) still satisfies this equation. To see this,
set $c' = c$ and $l' = l$ and check whether the budget constraint holds. Doing
this yields

$$c + w'l = w' - (w' - w)(1 - l).$$

Note that $c = w(1 - l)$, by the old budget constraint. Hence,

$$w(1 - l) + w'l = w' - w' + w'l + w(1 - l).$$

Thus, the left- and right-hand sides are equal.

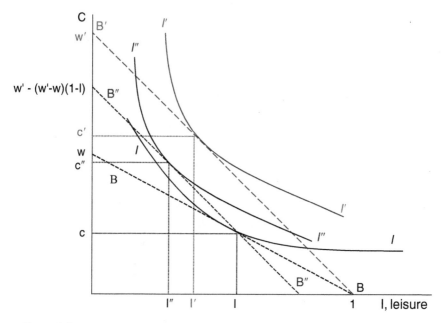

Figure 1.4
Substitution and income effects.

Example 1 (Compensating differential) *Wyle West lives in Denver. His employer is transferring him to New York City. Wyle currently spends $1,000 a month in rent, and $500 on food and entertainment. Rents are twice as high in NYC while food and entertainment are 1.3 times higher. So, his employer aims to give him an extra $$(2.0 - 1.0) \times 1,000 + (1.3 - 1.0) \times 500 = \$1,150$$ in salary a month. With this, he can still buy his old consumption bundle of housing plus food and entertainment. Wyle West will be better off, though, because he aims to live in a smaller place and spend more on food and entertainment, since the relative price of the latter is lower in NYC by a factor of $1.3/2.0 = 0.65$. That is, when facing the new NYC prices, Wyle West can still maintain his old spending habits, if he wishes, and hence will change them only if it makes him better off, which it will.*

The move from (l, c) to (l', c') can now be broken down into two steps, namely, from (l, c) to (l'', c'') and then from (l'', c'') to (l', c'). First, consider the move from point (l, c) to (l'', c''), which is associated with a shift in the budget line from BB to $B''B''$. This is a Slutsky compensated change in the real wage rate from w to w', because it reduces the person's income as much as possible while still allowing them to consume their old leisure-consumption bundle. The individual will move to the leisure-consumption bundle (l'', c'') on the new indifference curve $I''I''$. Observe that leisure falls and consumption rises with the compensated increase in the wage from w to w'. The relative price of leisure has risen so people consume less of it and switch to consumption. This is the *substitution effect* connected with a rise in the wage.

Second, move the budget constraint from $B''B''$ to $B'B$. This is connected with a rise in both consumption and leisure from (l'', c'') to (l', c'). This rise occurs when both consumption and leisure are *normal* goods. This is the *income effect* linked with the rise in the real wage. Therefore:

(1) Both the substitution and income effects associated with an increase in the real wage will cause consumption to rise. Hence, consumption will unambiguously move up with an increase in the real wage.

(2) The substitution effect causes leisure to fall, while the income effect leads it to rise.

(3) So, depending on which effect dominates, leisure may fall or rise. As will soon be seen, this depends on the form of the utility function, and hence the shape of the indifference curves, assumed for consumption and leisure.

2.2 Mathematical Formulations

The above diagrammatic analysis of the consumption-leisure choice in the unisex single model is now redone using mathematics. It's important to be able to go back and forth between economic intuition, which often derives from diagrams, and the mathematics of economics.

In modern economics tastes are specified by a utility function. Let

$U(c)$

represent the utility function for consumption. It gives the level of happiness, $U(c)$, that a person realizes, if they consume the amount, c. Some typical assumptions imposed on a utility function are:

Assumption 1 *U maps the nonnegative reals into the reals.* $U : \mathcal{R}_+ \to \mathcal{R}$. Consumption must always be nonnegative, but utility can be negative.

Assumption 2 *U is strictly increasing so that* $U_1 \equiv dU/dc > 0$. Marginal utility, U_1, is positive, implying that an extra unit of consumption increases utility; i.e., more consumption is always better than less. Therefore, even if utility is negative, it will still be increasing in consumption.

Assumption 3 *U is strictly concave so that* $U_{11} \equiv d^2U/dc^2 = dU_1/dc < 0$. Marginal utility, U_1, decreases as consumption increases. So, as consumption rises, each extra increment generates less and less additional utility. This is called *diminishing marginal utility*.

Example 2 (Common utility functions) Here are some utility functions that are commonly used in macroeconomics. They satisfy the above properties.

(1) Logarithmic utility:

$U(c) = \ln c$.

Here $U_1 = 1/c > 0$ and $U_{11} = -1/c^2 < 0$. Observe how utility may be positive or negative depending on whether $c \gtrless 1$. Note that $\lim_{c \to 0} U(c) = -\infty$. A special case of this utility function is $U(c) = \ln(c + \mathfrak{c})$, where $\mathfrak{c}$ is a constant. This utility function is strictly increasing and strictly concave, but now $\lim_{c \to 0} U(c) = \ln(\mathfrak{c})$; that is, it has a finite lower bound. This property will prove useful when discussing the decision to purchase health insurance in chapter 5. Furthermore, $U_{11} = -1/(c + \mathfrak{c})^2 > -1/c^2$. This implies that, for any given value for c, marginal utility is declining at a slower rate. This feature will be important for explaining the secular decline in fertility in chapter 2.

(2) Isoelastic utility:

$U(c) = c^{1-\rho}/(1-\rho) - 1/(1-\rho)$, *for* $\rho \geq 0$.

Now, $U_1 = c^{-\rho} > 0$ and $U_{11} = -\rho c^{-\rho-1} < 0$. Again, utility can be positive or negative. It can be shown, using L'Hopital's rule, that this utility function converges to the logarithmic case as $\rho \to 1$.[2] The parameter ρ is called the coefficient of relative risk aversion *and plays an important role in both economics and finance. Often the constant term, $-1/(1-\rho)$, is dropped from this utility function. Utility is linear in c when $\rho = 0$. This utility function is also known as a* constant relative risk aversion (crra) *utility function.*

---

2. For those interested, before applying L'Hopital's rule, note that $U(c) = (c^{1-\rho} - 1)/(1-\rho)$.

(3) Exponential utility:

$U(c) = -e^{-\gamma c}/\gamma$, *with* $\gamma > 0$.

For this utility function, $U_1 = e^{-\gamma c} > 0$ *and* $U_{11} = -\gamma e^{-\gamma c} < 0$. *Utility is always negative.*

(4) Quadratic utility:

$U(c) = \alpha c - \beta c^2/2$, *for* $\alpha, \beta > 0$ *and* $c < \alpha/\beta$.

In this case, $U_1 = \alpha - \beta c \gtreqless 0$ *and* $U_{11} = -\beta < 0$. *The quadratic utility function declines when* $c > \alpha/\beta$. *Hence, it is valid only over the domain for c where utility rises with c or for* $c < \alpha/\beta$. *The quadratic utility function subsumes the case of the linear function as a special case, a fact that can be seen by setting* $\beta = 0$. *A linear utility function is just concave, not strictly concave. A person with a linear utility function is often referred to as being* risk neutral. *The quadratic utility function is convenient to use when there is uncertainty in the analysis (as will be discussed in chapter 2). Its linear form is also employed in various places.*

Likewise, let

$V(l)$

represent the utility function for leisure. It returns the level of happiness, $V(l)$, that a worker realizes if they enjoy the fraction of time l in leisure. Leisure can be thought of as a good like consumption, so the properties for V are strictly analogous to those imposed on U. Now, as will be seen next, the relative strength of the income and substitution effects on leisure, resulting from a change in the wage rate, depends on the form of the utility functions assumed for consumption and leisure.

2.2.1   The unisex single model with logarithmic utility—equal income and substitution effects   Assume that the person's utility functions for consumption and leisure are logarithmic.[3] Specifically, suppose that the person's utility function is given by

$$u = \theta \ln(c) + (1 - \theta) \ln l, \text{ for } 0 < \theta < 1. \tag{2.1}$$

---

3. The logarithmic utility function gives the same solution for consumption and labor as the Cobb-Douglas utility function $u = c^\theta l^{1-\theta}$. This occurs because the logarithmic utility function is a strictly increasing monotonic transformation of the Cobb-Douglas one. In particular, just take the logarithm of the Cobb-Douglas utility function. Taking a strictly increasing monotonic transformation of the objective function in a maximization problem does not change its solution.

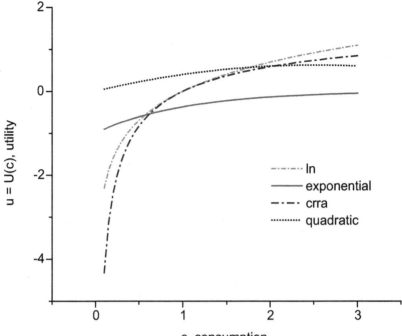

Figure 1.5
Utility functions: logarithmic (ln); isoelastic or crra ($\rho = 1.5$); exponential ($\gamma = 1$); quadratic ($\alpha = 0.5, \beta = 0.2$). Note how the quadratic utility function declines when $c > \alpha/\beta$. Hence, it is only good for $c < \alpha/\beta = 2.5$.

There will be indifference curves associated with this utility function. The above utility function is increasing in both c and l because the ln function is increasing. Thus, in any indifference curve diagram associated with this utility function, as one moves outward along a ray from the origin, one must travel through indifference curves that are associated with higher and higher levels of utility. Therefore, Property 1 will be satisfied. Additionally, each (l, c) combination is associated with a *unique* level of utility, u. Therefore, the indifference curves cannot cross, which is Property 2. Properties 3 and 4 will be verified shortly.

The individual's goal in life is to maximize utility subject to their budget constraint. The formal maximization problem is

$$\max_{c,l}\{\theta \ln(c) + (1 - \theta) \ln l\},$$

subject to

$$c = w(1 - l).$$

By substituting the budget constraint into the objective function to solve out for consumption, c, a maximization problem in just leisure, l, can be obtained. In particular, the recast problem is

$$\max_{l} \{\theta \ln(w(1 - l)) + (1 - \theta) \ln l\}.$$

At a maximum the first-order condition is zero, while the second-order condition is negative. (See the Mathematical Appendix for a brief discussion of maximization problems.)

The first-order condition associated with this maximization problem is

$$-\theta \frac{w}{w(1-l)} + (1 - \theta)\frac{1}{l} = 0, \tag{2.2}$$

which can be rewritten as

$$\theta \underbrace{\frac{1}{w(1-l)} \times w}_{\text{MU}_c} = \underbrace{(1 - \theta)\frac{1}{l}}_{\text{MU}_l}.$$

The left-hand side is the marginal benefit from working. By working an extra unit of time, the person earns w in income that is used to increase consumption, c. An extra unit of consumption will in turn increase utility by θ/c, which is the marginal utility of consumption denoted by MU_c. By using the person's budget constraint, it follows that $\text{MU}_c = \theta/c = \theta/[w(1-l)]$. The right-hand side is the marginal cost of working. This is the marginal utility of the foregone leisure, or $\text{MU}_l = (1 - \theta)1/l$.

The first-order condition simplifies to

$$\frac{l}{(1-l)} = \frac{(1 - \theta)}{\theta},$$

so that

$$l = 1 - \theta.$$

This solution for the unisex single model with logarithmic utility provides an important benchmark for comparison with the solution for the model of married female labor supply, presented later on. There are two key features to take away from the benchmark solution for leisure, l, or equivalently work, $1 - l$:

(1) Leisure, l, and hours worked, $1 - l$, do not depend on the wage rate, w. This is because the income and substitution effects from a change in wages exactly offset each other so that a change in wages has no impact on leisure or hours worked.

(2) The person always works, since $0 < 1 - l = \theta < 1$, and always enjoys some leisure, as $0 < l = 1 - \theta < 1$. That is, an interior solution to the above maximization problem always transpires. This occurs because as work effort and, hence, consumption go to zero, the marginal utility of consumption, $\theta/c = \theta/[w(1 - l)]$, goes to infinity. Therefore, the individual will always like to work at least a little bit. Similarly, as leisure goes to zero, the marginal utility of leisure, $(1 - \theta)/l$, goes to infinity. Consequently, the person will always want to enjoy a little bit of leisure. That is, the corner solutions $l = 0$ and $l = 1$ will never hold.

The second-order condition for a maximum requires that

$$\frac{d^2u}{dl^2} < 0,$$

where $u = \theta \ln(w(1 - l)) + (1 - \theta) \ln l$. Since the objective function is strictly concave in l, the second-order condition for a maximum is automatically satisfied. The objective function is strictly concave in l, because the utility functions for both consumption and leisure are strictly concave; i.e., both $\theta \ln(w(1 - l))$ and $(1 - \theta) \ln l$ are strictly concave in l. The sum of two strictly concave functions is strictly concave. It is easy to check directly, though, that

$$\frac{d^2u}{dl^2} = -\theta \frac{1}{(1-l)^2} - (1-\theta)\frac{1}{l^2} < 0.$$

What is the connection between the solution to the above maximization problem and figure 1.2? Note that the diagram sets the marginal rate of substitution of leisure for consumption, MRS, equal to the wage rate, w. To compute the marginal rate of substitution of leisure for consumption take the *total differential* of the logarithmic utility function (2.1) to get

$$du = \underbrace{\frac{\theta}{c}}_{MU_c} dc + \underbrace{\frac{(1-\theta)}{l}}_{MU_l} dl,$$

where MU_c and MU_l connote the marginal utilities of consumption and leisure. (For more on the concept of a total differential see the Mathematical Appendix.) This gives the change in utility, du, in response to a small shift in consumption in the amount dc and a small movement in leisure of the magnitude dl. Along an indifference curve utility is held constant so that $du = 0$. Therefore,

$$0 = \frac{\theta}{c}dc + \frac{(1-\theta)}{l}dl.$$

This gives a formula for the marginal rate of substitution of l for c:

$$\frac{dc}{dl}\bigg|_{\text{utility constant}} = -\frac{(1-\theta)}{\theta}\frac{c}{l} = -MRS = -\frac{MU_l}{MU_c} < 0. \tag{2.3}$$

Given the negative sign of the above expression, it is clear that the indifference curve slopes downward. So, Property 3 holds. Note that the marginal rate of substitution of leisure for consumption, $MRS = [(1-\theta)/\theta](c/l)$, is increasing in c and decreasing in l, as was conjectured. More formally,

$$\frac{d^2c}{dl^2}\bigg|_{\text{utility constant}} = -\frac{dMRS}{dl} = -\frac{(1-\theta)}{\theta}\frac{1}{l}\frac{dc}{dl}\bigg|_{\text{utility constant}} + \frac{(1-\theta)}{\theta}\frac{c}{l^2}$$

$$= \left[\frac{(1-\theta)}{\theta}\right]^2\frac{c}{l^2} + \frac{(1-\theta)}{\theta}\frac{c}{l^2} > 0 \text{ [using (2.3)]}.$$

This implies that the indifference curve satisfies Property 4.

Now, the equilibrium in figure 1.2 states that the marginal rate of substitution, MRS, should be set equal to the wage, w. Thus,

$$\frac{(1-\theta)}{\theta}\frac{c}{l} = w.$$

But, this is the first-order condition (2.2). To see this, set $c = w(1-l)$ in (2.2) and rearrange it to get the above equation. Thus, the two approaches coincide.

Observe that if one multiplies consumption by a factor of $\lambda > 1$, while holding leisure constant, then the marginal rate of substitution of leisure for consumption will also rise by that same factor. Now, turn to figure 1.6. Let wages rise by a factor of λ from w to $w' = \lambda w$, where $\lambda > 1$. As can been seen, leisure does not change as a result. This transpires because the slope of the indifference curve $I'I'$ at the point $(l, \lambda c)$ is exactly λ times higher (in absolute value) than the slope of the indifference curve II at the point (l, c). This fact

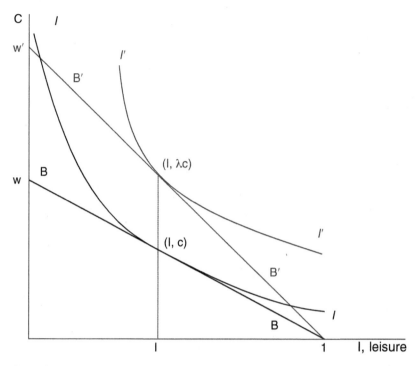

Figure 1.6
The impact of an increase in the real wage from w to w' when preferences are logarithmic.

follows from (2.3). Therefore, the indifference curve $I'I'$ will be tangent to the new budget line $B'B$ at the point $(l, \lambda c)$. This occurs because when l is held fixed, consumption will rise by λ, because $c' = w'(1-l) = \lambda w(1-l)$. Check for yourself that the income and substitution effects must exactly cancel out.

A heuristic derivation of the income and substitution effects of a change in wages will now be provided. By definition, the change in leisure due to a change in wages can be decomposed into substitution and income effects such that

$$\frac{dl}{dw} = \frac{dl}{dw}\bigg|_{\text{SUBST}} + \frac{dl}{dw}\bigg|_{\text{INCOME}}.$$

From the solution for leisure, it is immediately apparent that

$$\frac{dl}{dw} = 0,$$

so that

$$\frac{dl}{dw}\bigg|_{\text{SUBST}} = -\frac{dl}{dw}\bigg|_{\text{INCOME}}.$$

As a first step in calculating the income effect, give the person some income in the arbitrary lump-sum amount i. Their budget constraint would then appear as $c = w(1-l) + i$. Redoing the maximization problem gives the following solution for leisure:

$$l = (1-\theta)[1 + i/w],$$

implying

$$\frac{dl}{di} = \frac{(1-\theta)}{w} > 0.$$

The higher a person's income, i, is, the more leisure, l, they will enjoy. The second step involves computing the actual change in income, di, that will be associated with the change in wages, dw. If the person held hours worked, $1-l$, constant, then the resulting change in income would be $di = (1-l)dw$. Therefore, the positive income effect associated with a small increase in wages is

$$\frac{di}{dw} = 1 - l.$$

(The required Slutsky compensation, di, associated with a small change in wages, dw, is given by $-(1-l)dw$. After undertaking this negative compensation, the person could still keep their old (l, c) combination.) The income effect on leisure arising from an increase in wages is now easily seen to be

$$\frac{dl}{dw}\bigg|_{\text{INCOME}} = \frac{dl}{di}\frac{di}{dw} = \frac{(1-\theta)}{w}(1-l).$$

The substitution effect is just the negative of this.

2.2.2 The unisex single model with a zero-income effect utility function

For another example, consider the utility function

$$u = c - \frac{(1-l)^{1+\theta}}{1+\theta}, \text{ with } \theta > 0.$$

This function is quasilinear; that is, it is linear in c and nonlinear in l. Essentially, the disutility of working $(1-l)^{1+\theta}/(1+\theta)$ is being measured in consumption units. In this situation the individual's maximization problem will appear as

$$\max_{c,l} \left\{ c - \frac{(1-l)^{1+\theta}}{1+\theta} \right\},$$

subject to

$$c = w(1-l).$$

Again, solve out for consumption in the objective function to obtain

$$\max_{l \geq 0} \left\{ w(1-l) - \frac{(1-l)^{1+\theta}}{1+\theta} \right\}.$$

Undertaking the implied maximization (while assuming an interior solution) gives the first-order condition below:

$$w = (1-l)^{\theta},$$

so that

$$1 - l = w^{1/\theta}.$$

Observe that work effort, $1 - l$, is increasing in the wage rate, w. The term $1/\theta$ measures the elasticity of hours worked, $1 - l$, with respect to the wage rate, w. Strictly speaking the above solution is valid only when $l \geq 0$ or when $1 - l \leq 1$; i.e., the person cannot consume a negative amount of leisure or work more than their time endowment, 1. So, it is better to write the solution as

$$1 - l = \begin{cases} w^{1/\theta} > 0, & \text{when } w^{1/\theta} \leq 1; \\ 1, & \text{otherwise.}^4 \end{cases}$$

A person will *always* do *some* work, because $1 - l > 0$. The individual will consume zero leisure when the wage rate is equal to or greater than 1; i.e., $1 - l = 1$ when $w^{1/\theta} \geq 1$, or equivalently $l = 0$ when $w \geq 1$.

Definition 1 (Elasticity of Labor Supply) *The elasticity of hours worked with respect to wages measures the percentage change in hours worked in response to a percentage increase in wages. It is defined as*

$$\frac{\% \, \Delta \text{ in } 1 - l}{\% \, \Delta \text{ in } w} = \frac{w}{1-l} \frac{d(1-l)}{dw} = \frac{d\ln(1-l)}{d\ln w}.$$

4. Often this utility function is expressed as $c - h^{1+\theta}/(1+\theta)$, where h is hours worked. Usually an upper bound on hours worked is not imposed, so that h is allowed to be any nonnegative number.

Example 3 (Elasticity of labor supply)   *Let* $u = c - (1-l)^{1+\theta}/(1+\theta)$, *as assumed.*
It's clear that

$$\frac{d(1-l)}{dw} = \frac{1}{\theta} w^{1/\theta-1}.$$

Therefore, the elasticity of labor supply is

$$\frac{w}{1-l}\frac{d(1-l)}{dw} = \frac{1}{1-l}\frac{1}{\theta}w^{1/\theta} = \frac{1}{\theta}.$$

This utility function has no income effect on leisure. To see this, note that the marginal rate of substitution of leisure for consumption is

$$\mathrm{MRS} = -\frac{dc}{dl}\big|_{\text{utility constant}} = \frac{\mathrm{MU}_l}{\mathrm{MU}_c} = (1-l)^\theta.$$

The key point is that the marginal rate of substitution, MRS, does *not* depend on consumption, c. Hence, along a vertical line, the slope of the indifference curves does not change (see figure 1.7). When $l=0$, then $dc/dl|_{\text{utility constant}} = -1$, and when $l=1$, then $dc/dl|_{\text{utility constant}} = -0$. It is easy to tell from the diagram that since the slope of the budget constraint is always negative (or <0), the person will always do some work ($l<1$). Likewise, the person will work full time ($l=0$) when the wage rate is greater than or equal to 1. Now, consider an increase in income shown by a parallel shift in the budget line from BB to $B'B'$. Leisure remains constant, while consumption increases from c to c'.

Another way to see this is to give the individual some income in a lump-sum amount, say i. Their budget constraint would now read $c = w(1-l) + i$. The maximization problem would appear as

$$\max_l \left\{ w(1-l) + i - \frac{(1-l)^{1+\theta}}{1+\theta} \right\}.$$

Check for yourself that the solution for labor supply is still given by

$$1 - l = w^{1/\theta}.$$

Hence, the amount of lump-sum income, i, does not affect the person's labor supply. This implies the income effect must be zero.

2.2.3   The unisex single model with Leontief utility—no substitution effect

For the last example, suppose that the individual's utility function is

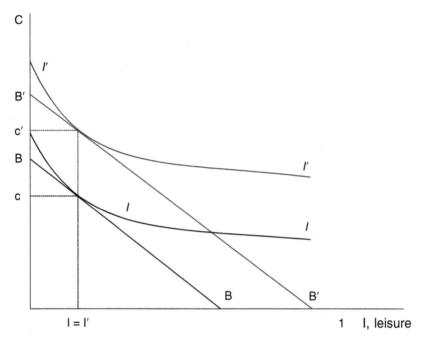

Figure 1.7
Illustration of the indifference curves associated with the zero-income effect utility function. Note that the slopes of the indifference curves are -1 when $l=0$ and 0 when $l=1$. Along any vertical line the slopes of the indifference curves are the same.

given by

$$u = \min\{\theta c, (1-\theta)l\}, \quad \text{for } 0 < \theta < 1.$$

This utility function is not amenable to calculus. It implies the individual will enjoy consumption and leisure in fixed proportions. Specifically, it is easy to deduce that the person will always choose c and l so that $\theta c = (1-\theta)l$; otherwise, the individual will be wasting some resources on either consumption or leisure. This implies

$$c = \frac{1-\theta}{\theta} l.$$

Substituting this solution for c into the person's budget constraint gives

$$c = \frac{1-\theta}{\theta} l = w(1-l),$$

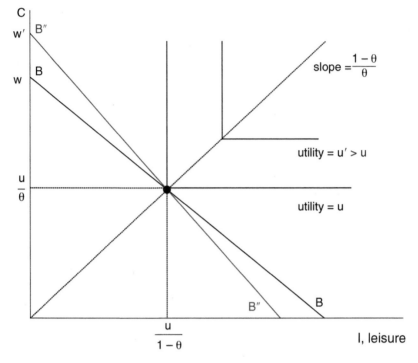

Figure 1.8
Illustration of the zero-substitution effect utility function.

so that

$$l = \frac{w}{(1-\theta)/\theta + w}.$$

As can be seen, a rise in the real wage will lead to an increase in leisure. Therefore, the income effect must dominate the substitution effect. In fact, there is no substitution effect.

To see this, examine figure 1.8, which portrays an equilibrium for the individual. Note that the person's indifference curves are L-shaped. To obtain the utility level u requires u/θ units of consumption *and* $u/(1-\theta)$ units of leisure. Giving the individual more consumption (leisure) while holding leisure (consumption) fixed would not increase their utility. The kinks in all of the L-shaped indifference curves lie somewhere on the dotted ray emanating from the origin. Imagine rotating the budget line from BB to $B''B''$. This would happen when there is a Slutsky compensated shift in the real wage. Consumption and leisure do not change.

3 Married Female Labor Supply

It's time to move away from the unisex single model of labor supply. Two models of a married woman's labor supply will be considered: one in which labor supply can be adjusted along the intensive margin and another in which labor supply moves along the extensive margin. Loosely speaking, the intensive margin allows the household freely to choose the time the woman will spend working in the market. In the intensive model of labor supply, all households are the same. Each household has a continuous choice problem where it selects the woman's hours of work. The focus of this analysis is on the number of hours a married woman will spend working. The extensive margin model of labor supply focuses on whether or not the women should enter the workforce, a discrete choice problem, while fixing the time spent at work at say, 40 hours a week. The extensive margin is convenient when analyzing some sort of heterogeneity across households. In this framework households differ by how much they value the option of the woman staying at home. The analysis spotlights the fraction of married woman that will work.

The primary goal of this chapter is to explain the long-run rise in married female labor-force participation. This section is a stepping stone toward that end. The analysis presumes a traditional married family in which the man works full time in the market, and the household makes a joint decision about the woman's participation in the labor market. This has been the predominant family structure in American history. Of course, in modern times some couples cohabit while not being married. Others are married and are of the same sex. Less is known about these more modern forms of households, both empirically and theoretically. Analyzing all three possible household types and their evolution, as is done for the traditional family in this chapter and the later one on marriage, would be far too complicated. It would even involve voyaging over some uncharted waters in economics. However, section 9 of this chapter, on the division of labor within the family, is germane for these types of households as well.

3.1 Married Female Labor Supply along the Intensive Margin

Imagine a married couple. Each person has one unit of time. Suppose that the husband always works (this assumption is dropped in section 9). He spends the fixed amount $\bar{h}_m$ in the market at the wage rate w_m. The couple collectively decides how much the wife should work. Denote her hours in the market by h_f. Her wage rate is w_f. The ratio of women's to men's wages is called the *gender wage gap*. The gender wage gap is defined by $\phi \equiv w_f / w_m$. Exactly what determines the gender wage gap is unclear. A discussion of the this is postponed to section 7. Suppose that consumption and leisure in the household are public goods. That is, each party collectively enjoys the household's consumption, c, and each other's leisure, $1 - h_f$ and $1 - \bar{h}_m$. So, while the man may enjoy his guitar and the woman her piano, the husband shares equally on his wife's happiness from the piano and vice versa. Specifically, express the household's preferences as

$$u = \theta \ln c + (1 - \theta) \ln(1 - h_f) + (1 - \theta) \ln(1 - \bar{h}_m), \quad \text{with} \quad 0 < \theta < 1. \qquad (3.1)$$

The couple's budget constraint reads

$$c = w_m \bar{h}_m + w_f h_f.$$

From the earlier analysis, it is apparent that the couple's maximization problem can be written as

$$\max_{h_f \geq 0} \left\{ \theta \ln(\underbrace{w_m \bar{h}_m + w_f h_f}_{c}) + (1 - \theta) \ln(1 - h_f) \right\}.$$

The term $(1 - \theta) \ln(1 - \bar{h}_m)$ can safely be dropped from the objective function because it represents a constant, which will vanish upon differentiation. This optimization problem has two solutions, namely an interior solution and a corner solution. (See the Mathematical Appendix for a discussion of interior and corner solutions). The first-order condition for the interior solution is

$$\theta \underbrace{\frac{w_f}{w_m \bar{h}_m + w_f h_f}}_{\text{MB}} = (1 - \theta) \underbrace{\frac{1}{1 - h_f}}_{\text{MC}}. \qquad (3.2)$$

The left-hand side of the above equation is the marginal benefit, MB, from the woman working. By spending an extra unit of time in the market, the woman will earn w_f, which can be used for consumption. An extra unit of consumption is worth $\theta/(w_m \bar{h}_m + w_f h_f)$ in extra utility; that is, $MU_c = \theta/c = \theta/(w_m \bar{h}_m + w_f h_f)$. The right-hand side is the marginal cost, MC, of working an extra unit, which is the marginal utility of the woman's leisure; that is $MU_{1-h_f} = (1 - \theta)/(1 - h_f)$. Thus, the equation states

$MU_c \times w_f = MU_{1-h_f}$. Solving the above first-order condition for h_f gives

$$h_f = \theta - (1-\theta)\frac{w_m}{w_f}\overline{h}_m \quad \text{(interior solution)}.$$

The woman's labor supply, h_f, can be negative in the above equation, which would be a nonsensical answer. It is easy to deduce that this happens when the woman's wage is low—in particular, when $w_f \leq \frac{(1-\theta)}{\theta}w_m\overline{h}_m$. In this situation, the true solution is

$$h_f = 0 \quad \text{(corner solution)}.$$

The corner solution happens when MB < MC in (3.2), evaluated at $h_f = 0$. In this situation, it is not worth the woman in the household going to work because the marginal benefit from increasing her hours from zero is less than the marginal cost.

So, the complete solution for h_f is

$$h_f = \begin{cases} \theta - (1-\theta)\frac{w_m}{w_f}\overline{h}_m \text{ (work)}, & \text{if } w_f \geq \frac{(1-\theta)}{\theta}w_m\overline{h}_m \text{ (interior solution)}; \\ 0 \text{ (don't work)}, & \text{if } w_f < \frac{(1-\theta)}{\theta}w_m\overline{h}_m \text{ (corner solution)}. \end{cases}$$

This labor supply function has four interesting properties:

(1) A married working woman's labor supply, h_f, is increasing in her own wage, w_f. This can be seen from the upper branch of the equation. Compare this with the unisex single model with logarithmic preferences. There, wages had no effect on labor supply. Observe that the unisex single model can be recovered from the above formula by setting $\overline{h}_m = 0$. Then, effectively, there is no man in the household. In this case, $h_f = \theta$, a constant that is invariant to the wages.

(2) A married working woman's labor supply, h_f, is decreasing in relation to her husband's wage, w_m. This represents an income effect. As the husband's wage increases, the household becomes richer, and the extra income is used to increase both household consumption and the wife's leisure.

(3) The woman may not work, implying $h_f = 0$. This is shown by the lower branch. This branch occurs when the marginal benefit from working at $h_f = 0$ (zero labor supply) is less than the marginal cost, or when $MU_c \times w_f = \theta w_f/(w_m\overline{h}_m) < (1-\theta) = MU_{1-h_f}$. This condition can be rewritten as $w_f < [(1-\theta)/\theta]w_m\overline{h}_m$. That is, the woman will stay at home when her wage is low (at least relative to her husband's). She will also stay at

home when the weight on her leisure in utility is high relative to the weight on consumption, or when $(1-\theta)/\theta$ is large. At the corner solution the woman's wage can rise without an accompanying increase in her labor supply.

(4) An equal percentage increase in both wages has no effect on labor supply. To see this, let $w_f' = \lambda w_f$ and $w_m' = \lambda w_m$, where $\lambda > 1$ is the gross growth rate in wages. Now, $w_m'/w_f' = (\lambda w_m)/(\lambda w_f) = w_m/w_f$, so labor supply will remain the same. This is reminiscent of the result in the unisex single model where an increase in wages had no impact on labor supply.

(5) Observe that one could write

$$
h_f = \begin{cases} \theta - (1-\theta)\frac{1}{\phi}\bar{h}_m \text{ (work)}, & \text{if } \phi \geq \frac{(1-\theta)}{\theta}\bar{h}_m; \\ 0 \text{ (don't work)}, & \text{if } \phi < \frac{(1-\theta)}{\theta}\bar{h}_m, \end{cases}
$$

where again $\phi \equiv w_f/w_m$. So really only changes in the gender wage gap, ϕ, influence married female labor supply, h_f.

Consider the case where an interior solution holds. The first-order condition (3.2) can be rearranged to give

$$
\text{MRS} = -\frac{dc}{d(1-h_f)}\Big|_{\text{utility constant}} = \frac{(1-\theta)}{\theta}\frac{w_m\bar{h}_m + w_f h_f}{1-h_f}
$$

$$
= \frac{(1-\theta)}{\theta}\frac{c}{1-h_f} = \text{MU}_{1-h_f}/\text{MU}_c = w_f.
$$

The left-hand side is the marginal rate of substitution of the woman's leisure for household consumption. This should be set equal to the woman's wage, the right-hand side. Figure 1.9 portrays the situation graphically. The marginal rate of substitution measures the amount of consumption that the household would be willing to trade for an extra unit of the woman's leisure. It is the slope of the indifference curve between consumption and leisure. (Actually, it is the negative of the slope of the indifference curve.) An increase in the woman's wage from w_f to $w_f' = \lambda w_f$ rotates the budget line from BB to BB'. Along the vertical line at $1 - h_f$ the slope of an indifference curve rises by *less* than the factor λ. This is easy to see from the above formula for the MRS. While $w_f h_f$ increases by λ, the term $w_m\bar{h}_m$ remains constant; hence the term $c = w_m\bar{h}_m + w_f h_f$ rises by less by λ. Therefore, the new equilibrium must lie to the left of the square. The woman's leisure drops as a consequence, while the household's consumption moves up.

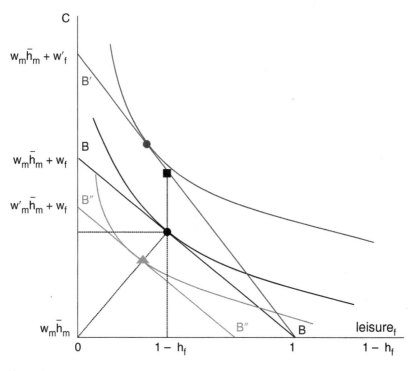

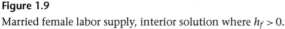

Figure 1.9
Married female labor supply, interior solution where $h_f > 0$.

In contrast, a drop in the male's wage, w_m, leads to the budget constraint shifting down in parallel fashion from BB to $B''B''$. This acts as a negative income effect. The new equilibrium is shown by the triangle. It must lie on a ray from the origin that goes through the old equilibrium point. This occurs because the ratio of c to $(1 - h_f)$ remains constant, because w_f has not changed. To see this, note that given the form of the utility function, $[(1 - \theta)/\theta]c/(1 - h_f) = MRS = w_f$. This results in both the woman's leisure and the household's consumption falling.

The corner solution situation where $h_f = 0$ is shown in figure 1.10. At the low wage rate w_f, the household decides that the woman should not work. The indifference curve II goes through the budget line BB at the corner point K, where $h_f = 0$ (so that $1 - h_f = 1$). The indifference curve is steeper than the budget line at this point. What does this say? It indicates that the amount of consumption that the household is to willing to give up for an

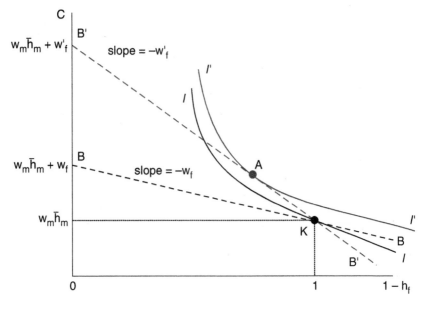
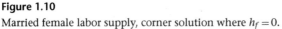

Figure 1.10
Married female labor supply, corner solution where $h_f = 0$.

extra unit of the wife's leisure is bigger than w_f, or that MRS $> w_f$. Thus, if the household cut the woman's labor effort by a small amount, the utility loss from the drop in household consumption would be less than the utility gain from the increase in woman's leisure. Of course they can't execute such a trade because the wife is already at her maximum amount of leisure, $1 - h_f = 1$, or equivalently is not working. In terms of the first-order condition (3.2), the left-hand side is less that the right-hand side, when evaluated at $h_f = 0$, so that $\theta w_f / (w_m \bar{h}_m) < (1 - \theta)$. This just says that the marginal benefit of working is less than the marginal cost, or $MU_c \times w_f < MU_{1-h_f}$. The woman does work at the higher wage rate w'_f. Here the budget line $B'B'$ is tangent to the indifference curve $I'I'$ at an interior point, A.[5]

5. It is worth comparing figures 1.9 and 1.10. In figure 1.9 there is an indifference curve (not shown) going through the point $(1, w_m \bar{h}_m)$. At this point, the slope of the indifference curve is less than the slope of the budget constraint, BB. Therefore, the amount of consumption the household requires in order to give up a unit of the wife's leisure is less than the wage rate. Hence, the woman should work. This is analogous to the situation in figure 1.10. Observe that the slope of the indifference curve $I'I'$ is less than the slope of budget line $B'B'$ at the point K in figure 1.10. Note

3.2 Married Female Labor Supply along the Extensive Margin

In the previous analysis a rise in female labor-force participation is attained by enticing all women to work more hours in the market. Suppose now that the workweek by a person is held fixed, say at 40 hours a week. The decision for an individual is whether to work or not; that is, it is a *discrete* choice. The married female labor-force participation decision is made along the *extensive margin*. An increase in female labor-force participation can occur only by persuading more women to leave home and enter the market. Return to the married couple's problem. Imagine that the woman must spend $\bar{h}_f$ hours on the job, if she works. That is, h_f is contained in the two-point set, $\mathcal{H}$, so that

$$h_f \in \mathcal{H} \equiv \left\{ \underbrace{0}_{\text{don't work}}, \underbrace{\bar{h}_f}_{\text{work}} \right\}.$$

As before, the man has a fixed workweek, $\bar{h}_m$ (again, this assumption is relaxed in section 9).

Express the household's preferences as

$$u = \theta \ln c + (1-\theta)\lambda \ln(1 - h_f) + (1-\theta)\ln(1 - \bar{h}_m), \text{ with } 0 < \theta < 1.$$

The variable $\lambda \geq 0$ denotes the value that the couple places on the woman's time spent at home. It differs across households. Some households value the woman's time at home more than others. They will have a high value for λ. Different households may hold different opinions about the role of the woman at home, especially when there are dependent children. In particular, assume that λ is distributed across members of society according to some distribution function (more on that momentarily). The couple's budget constraint reads

$$c = \begin{cases} w_m \bar{h}_m + w_f \bar{h}_f, & \text{work;} \\ w_m \bar{h}_m, & \text{don't work.} \end{cases}$$

Should the woman work or not?

To answer this question, compute the utility associated with each option. If the woman works, then the utility that the household will realize is

$$u_{\text{work}} = \theta \ln(w_m \bar{h}_m + w_f \bar{h}_f) + (1-\theta)\lambda \ln(1 - \bar{h}_f) + (1-\theta)\ln(1 - \bar{h}_m). \tag{3.3}$$

---

that the vertical axis in figure 1.9 starts off at $w\bar{h}_m$ so, unlike figure 1.10, the bottom of the diagram has been cropped off.

Alternatively, if she stays at home, then household utility will be given by

$$u_{\text{don't work}} = \theta \ln(w_m \overline{h}_m) + (1 - \theta) \ln(1 - \overline{h}_m),\tag{3.4}$$

as $\ln(1) = 0$. The couple will pick the option that maximizes the household's utility. Therefore,

$$h_f = \begin{cases} \overline{h}_f \text{ (work)}, & \text{if } \theta \ln(w_m \overline{h}_m + w_f \overline{h}_f) \\ & \quad + (1 - \theta)\lambda \ln(1 - \overline{h}_f) \geq \theta \ln(w_m \overline{h}_m); \\ 0 \text{ (don't work)}, & \text{if } \theta \ln(w_m \overline{h}_m + w_f \overline{h}_f) \\ & \quad + (1 - \theta)\lambda \ln(1 - \overline{h}_f) < \theta \ln(w_m \overline{h}_m). \end{cases}$$

Observe that the term $(1 - \theta) \ln(1 - \overline{h}_m)$ cancels out when u_{work} is compared with $u_{\text{don't work}}$. (When $u_{\text{work}} = u_{\text{don't work}}$ the household doesn't care whether the woman works or not. So, in this situation, it doesn't really matter in which category the work decision is assigned.)

There will exist some λ that makes the household indifferent regarding the woman working or not working. Denote this value of λ by λ^*. This is called the *threshold value* for λ. It is obtained by setting $u_{\text{work}} = u_{\text{don't work}}$ and then solving for λ. This gives

$$\lambda^* = \frac{\theta}{(1 - \theta)} \frac{[\ln(w_m \overline{h}_m + w_f \overline{h}_f) - \ln(w_m \overline{h}_m)]}{-\ln(1 - \overline{h}_f)}$$

$$= \frac{\theta}{(1 - \theta)} \frac{\ln[1 + w_f \overline{h}_f / (w_m \overline{h}_m)]}{-\ln(1 - \overline{h}_f)} \geq 0,$$

where the second line follows from the fact that $\ln(A + B) - \ln(A) = \ln[(A + B)/A] = \ln(1 + B/A)$, where A and B are greater than zero. Clearly, a woman will work when $\lambda \leq \lambda^*$ and will stay at home when $\lambda > \lambda^*$. Hence, the decision about whether the woman should work or not can be recast as

$$h_f = \begin{cases} \overline{h}_f \text{ (work)}, & \text{if } \lambda \leq \lambda^*; \\ 0 \text{ (don't work)}, & \text{if } \lambda > \lambda^*. \end{cases}$$

The threshold value, λ^*, is increasing in w_f and decreasing in w_m. Last, $\lambda^* \in (0, \infty)$. If men earned nothing, so that $w_m = 0$, then all women would work, since $\lim_{w_m \to 0} \lambda^* = \infty$. Additionally, as men's wages grow without bound, *ceteris paribus*, no women would work; i.e., $\lim_{w_m \to \infty} \lambda^* = 0$. No women work when $w_f = 0$, and all women work as $w_f \to \infty$.

To characterize how many married women will work requires knowing how λ is spread across the population. Cumulative distribution functions are very useful for this purpose. Suppose that $\lambda \in [\psi, \infty)$ is distributed across the population according to the Pareto distribution, $1 - (\psi/\lambda)^\gamma$, with $\psi > 0$ and $\gamma > 1$. (See the Mathematical Appendix for a discussion of distribution functions and the definition of the Pareto distribution.) Observe that $1 - (\psi/\lambda)^\gamma$ increases monotonically from 0 to 1 as λ rises from its lower bound ψ to ∞. The number $1 - (\psi/\lambda^*)^\gamma$ is the fraction of households with a value of λ less than or equal to λ^*. The fraction of women working in society, μ, is then given by

$$\mu = \Pr[\lambda \le \lambda^*] = 1 - \left(\frac{\psi}{\lambda^*}\right)^\gamma.$$

Female labor-force participation, μ, is increasing in the threshold value, λ^*. In all households with a value for λ below λ^* the woman will work. These households have a (relatively) low value for the woman staying at home.

The model of married female labor supply along the extensive margin behaves remarkably similarly to the one along the intensive margin. The following four properties hold:

(1) The fraction of married women working, μ, is increasing in the woman's wage, w_f. To see this, note that λ^* is increasing in w_f and that μ is rising in λ^*. Therefore, an increase in women's wages leads to a rise in the threshold value of λ required for a woman to stay at home. Hence, fewer women will stay at home.

(2) The fraction of married women working, μ, is decreasing in the male wage, w_m. A rise in w_m lowers the threshold value, λ^*. A drop in the threshold value, λ^*, implies fewer women will be working.

(3) Not all women will work (at least when the husband earns some income or when $w_m > 0$). In particular, any woman with a λ greater than λ^* stays at home.

(4) An equal percentage increase in men's and women's wages has no effect on the fraction of women who work. Again, let $w_f' = \omega w_f$ and $w_m' = \omega w_m$, where $\omega > 1$ is the gross growth rate in wages. Such a change has no impact on the threshold value for λ^*, because $w_f'/w_m' = w_f/w_m$.

(5) The condition for the threshold value of λ can be rewritten as

$$\lambda^* = \frac{\theta}{(1-\theta)} \frac{\ln(1 + \phi \bar{h}_f/\bar{h}_m)}{-\ln(1 - \bar{h}_f)} \ge 0.$$

The only thing that matters is the gender wage gap, ϕ.

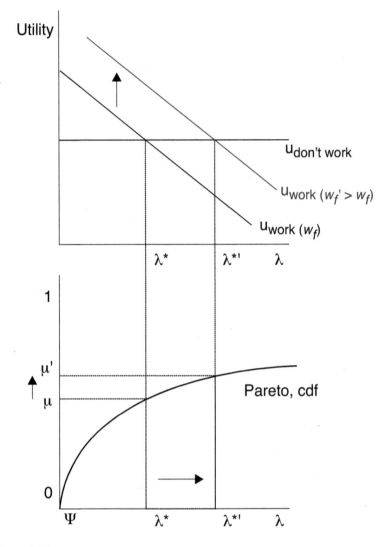

Figure 1.11
Married female labor supply along the extensive margin. The diagram also shows what happens when the wage rate for women rises from w_f to w_f'.

Figure 1.11 illustrates the situation graphically. The upper panel plots the utility levels for the household when the woman works and when she does not work as a function of λ. The "work" line decreases in the value that the household places on the woman staying at home, as measured by λ; recall (3.3) and note that $\ln(1 - \bar{h}_f) < 0$ because $1 - \bar{h}_f < 1$. The "don't work" line is horizontal because here utility is not actually a function of λ; see (3.4). The threshold value of λ or λ^* is determined by the intersection of these two lines. The bottom panel draws the cumulative distribution function (cdf) for the Pareto distribution. Given a threshold value for λ, it is then easy to calculate how many women will work as given by μ. A rise in women's wages from w_f to w_f' will cause the "work" line to move up. It is immediate that this leads to more women working, as shown by the movement from μ to μ'.

4 Calibrating Labor Supply Models

Economic theory comes alive when confronted with data. One method of matching economic models with data is *calibration*. Calibration often refers to adjusting an instrument, scientific or otherwise, so that it matches some known benchmarks. For example, a guitar can be tuned so that the A string has the Stuttgart pitch of 440Hz. After the guitar has been tuned (calibrated), it can be used to play songs in key with others. In economics the model is treated as an instrument, and its parameters can be adjusted so that it matches certain features in the data. After an economic model has been calibrated (tuned), it can be used to conduct policy analysis. Can the above models be made to fit some stylized facts computed from table 1.1? In particular, can the models mimic the average hours worked by a person over the postwar period? The answer is yes. The strategy is to reverse-engineer the weights on leisure (or equivalently consumption) to make the model replicate the fraction of time a person spends working. This is done first for the unisex single model and then for the model of married female labor supply along the intensive margin.

Example 4 (Unisex single model) *An individual has approximately 112 non-sleeping hours a week (16 × 7) available. Therefore, about 30 percent of a person's time is spent on total market work ($\simeq 100 \times 34/112$), and about 70 percent is time spent at home. Recall that the solution for leisure in the unisex single model is*

$l = 1 - \theta.$

Therefore, a good value to choose for the weight on the utility from leisure in the unisex single model is $1 - \theta = 0.70$, *so that* $\theta = 0.30$.

Example 5 (Married female labor supply, intensive margin) *On average over the postwar period a man spent about 45 hours a week on total market work. By contrast, a woman worked on average about 24 hours a week. Recall that the solution for married female labor supply is*

$$h_f = \theta - (1 - \theta)\frac{w_m}{w_f}\bar{h}_m.$$

Now, the above facts imply that a solution is desired where $h_f = 24/112 = 0.21$. *The facts also impose the restriction* $\bar{h}_m = 45/112 = 0.40$. *Assume the wage rate for a women is 70 percent of a man's (a reasonable number for the 1990s). Therefore,* $\phi = w_f/w_m = 0.70$, *so that* $w_m/w_f = 1/0.70$. *Solve the above equation for* θ. *It is easy to see that getting the observed facts requires setting the weight on the utility from consumption to 0.497:*

$$\theta = \frac{h_f + (w_m/w_f)\bar{h}_m}{1 + (w_m/w_f)\bar{h}_m} = \frac{0.21 + 0.40/0.70}{1 + 0.40/0.70} = 0.497. \tag{4.1}$$

Example 6 (Tax policy analysis) *The above calibrated model can be used to evaluate the impact of tax policy on married female labor supply. Suppose that the government taxes labor income to finance government expenditure, g. The household values this expenditure according to the strictly increasing, strictly concave utility function* $G(g)$; *i.e.,* $G_1 > 0$ *and* $G_{11} < 0$. *This function is simply added on to (3.1). Let* w_m *and* w_f *represent the before-tax wage rates for men and women and* τ_m *and* τ_f *denote the gender-specific tax rates on men's and women's labor income. The solution for* h_f *is now*

$$h_f = \theta - (1 - \theta)\frac{(1 - \tau_m)w_m}{(1 - \tau_f)w_f}\bar{h}_m,$$

where $(1 - \tau_m)w_m$ *and* $(1 - \tau_f)w_f$ *are the after-tax wage rates for men and women. Since government spending enters the household utility in an additive fashion, the level of g does not affect the household's first-order condition. If the government initially taxes the labor income of both men and women at 20 percent, then the calibration equation (4.1) will still hold because* $(1 - \tau_m)/(1 - \tau_f) = 1$, *so that once again* $\theta = 0.497$. *At these tax rates men earn more than women, both because their wage rate is higher and because they work more. The government proposes to increase the tax rate on men to 22 percent and to reduce the one on women to 13 percent, which they estimate will be revenue neutral. By using the above formula, while setting* $\theta = 0.497$, *one can deduce that the proposed policy will increase married female labor supply by 14 percent from 0.21 to 0.24. There is no income effect associated with the proposed tax change, because the policy is revenue neutral; i.e., the household will pay the same amount of taxes under both policies. Therefore, only the positive substitution effect from increasing the woman's after-tax wage rate operates. This works to increase married female labor supply.*

Another illustration of the calibration methodology is provided in section 8 of chapter 4.[6]

5 Household Production Theory

Margaret Reid (1934) was the first person to introduce the notion of household production into economics, in her Ph.D. dissertation (1931) at the University of Chicago. Household production theory treats the home as a small factory. The home employs inputs, intermediate goods and labor, to manufacture home goods. As will be seen, production in the home can be subject to technological progress, just like production in the market. According to Reid (1934, v),

The household is our most important economic institution. Yet economics of household production is a neglected field of study. With few exceptions the interest of economists has been concentrated on that part of our economic system which is organized on a price basis. The productive work of the household has been overlooked, even though more workers are engaged in it than any other single industry.

Reid (1934, 167) reported a value of housewives services of $15.3 billion in 1918 relative to a national income of $61 billion, which amounted to 25 percent of national income. Becker (1965) used modern microeconomic theory to formalize the notion of household production.

5.1 The Model Set-Up
Suppose that the household produces home goods, n, according to the following *Cobb-Douglas production function*

$$n = d^{1-\kappa} l^{\kappa}, \text{ with } 0 < \kappa < 1, \tag{5.1}$$

where d is the amount of intermediate goods used, and l represents the amount of household labor. The exponent κ measures the value of labor

6. See also the problems in chapters 1, 2, and 4.

in household production. The bigger it is, the more important labor is (relative to intermediate goods). To keep the analysis simple, let d be indivisible. Specifically, suppose that d lies in the two-point set, $\mathcal{D}$, so that $d \in \mathcal{D} \equiv \{\underline{\delta}, \overline{\delta}\}$, where $\underline{\delta} < \overline{\delta}$. Let p represent the price of intermediate goods. Assume that $p = 0$ when $d = \underline{\delta}$, and $p = q$ for $d = \overline{\delta} > \underline{\delta}$. Think about d as a proxy for the products used at home, such as cell phones, frozen foods, irons, microwaves, PCs, refrigerators, vacuum cleaners, washing machines, Tupperware, *inter alia*. The best technology, $\overline{\delta}$, has lots of these goods while the worst one, $\underline{\delta}$, has few ($\overline{\delta} > \underline{\delta}$). So, for a given amount of household labor, l, the best technology, $d = \overline{\delta}$, will produce more home goods, n, than the worst one, $d = \underline{\delta}$. The best technology costs more than the worst one ($q > 0$).

Endow the household with the following tastes:

$$u = \theta \ln c + (1 - \theta) \ln n, \quad \text{with } 0 < \theta < 1.$$

Tastes are distributed over the consumption of market goods, c, and home goods, n. Observe that leisure does not enter into the household's preferences. In this regard, the framework is a simple take on Becker (1965, 504), who felt that "although the social philosopher might have to define precisely the concept of leisure, the economist can reach all his traditional results as well as many more without introducing it at all."

Once again a man and a woman each has one unit of time. The man spends *all* of his time working in the market. The woman can split her time between house work, l_f, and working in the market, $1 - l_f$. The time the wife spends at home is valuable in a productive sense, in line with Reid (1934). The household's budget constraint appears as

$$c = \begin{cases} w_m + w_f(1 - l_f), & \text{if } d = \underline{\delta} \text{ (worst);} \\ w_m + w_f(1 - l_f) - q, & \text{if } d = \overline{\delta} \text{ (best).} \end{cases} \tag{5.2}$$

Assume that it is feasible for the household to purchase the best technology when both the man and woman work full time in the market.

Condition 1 (Affordability) $w_m + w_f - q > 0$.

The household has to pick how many market and home goods to consume, which technology to purchase, and the amount of housework the woman should do. Given the above setting, the household's maximization problem can be posed as

$$\max_{c, n, l_f, d \in \{\underline{\delta}, \overline{\delta}\}} \{\theta \ln c + (1 - \theta) \ln n\},$$

subject to the household production function (5.1) and the budget constraint (5.2). In the above problem, the household's budget constraint can be used to substitute out for c and, likewise, the production function for n. This eliminates c and n. It pays to break this problem down into two stages. In the first stage, the household decides how many hours the woman should work at home, l_f, given some level of intermediate goods, d. The household then decides which bundle of intermediate goods, $d \in \{\underline{\delta}, \overline{\delta}\}$, to buy; this is the second stage.

5.2 The Housework Decision, Stage 1

Now, suppose that the household picks the durable level d at the price p. The quantity-price bundle, (d, p), can be either $(\underline{\delta}, 0)$ or $(\overline{\delta}, q)$ corresponding to the worst and best technologies, respectively. Given this choice, the household's problem simplifies to

$$\underbrace{V(w_f, w_m, p, d)}_{\text{INDIRECT UTILITY}} = \max_{l_f} \{\theta \ln[w_m + w_f(1 - l_f) - p]$$

$$+ (1 - \theta)(1 - \kappa) \ln(d) + (1 - \theta)\kappa \ln(l_f)\}. \tag{5.3}$$

The function $V(w_f, w_m, p, d)$ is the household's *indirect utility function*. It gives the *maximal* level of utility that the household can realize when it faces the prices w_f, w_m and p and the stock of durables is d. It is predicated on the fact that the household picks l_f optimally. Intuitively, one would expect $V(w_f, w_m, p, d)$ to be decreasing in p. That is, the family is worse off the more expensive the household technology is. Likewise, one would think that $V(w_f, w_m, p, d)$ is increasing in d, because the household will be better off the larger is the bundle of intermediate goods that they are buying at the price p. Both of these properties will be established later. In a similar vein, $V(w_f, w_m, p, d)$ rises with both the woman's and the man's wages, w_f and w_m.

Undertaking the required maximization gives

$$\theta \underbrace{\frac{1}{w_m + w_f(1 - l_f) - p}}_{\text{MU}_c} \times w_f = \kappa(1 - \theta)\underbrace{\frac{1}{l_f}}_{\text{MU}_n \times \text{MP}_{l_f}}. \tag{5.4}$$

As before, the left-hand side represents the marginal benefit from the woman working an extra unit of the time in market. By increasing her time by a unit in the market, the woman will earn w_f. This can be used for consumption. An extra unit of consumption generates $\theta/c = \theta/[w_m + w_f(1 - l_f) - p]$ in extra utility. The right-hand side is the marginal cost. By taking a unit of time away from home production the output of home goods will drop by marginal product of labor at home, MP_{l_f}. It is easy to calculate that $\mathrm{MP}_{l_f} = \kappa d^{1-\kappa} l_f^{\kappa-1}$. The marginal utility of nonmarket goods, MU_n, is given by $\mathrm{MU}_n = (1 - \theta)/n = (1 - \theta)/(d^{1-\kappa} l_f^{\kappa})$. So, the marginal cost of taking a unit of time away from home is $\mathrm{MU}_n \times \mathrm{MP}_{l_f} = (1 - \theta)/(d^{1-\kappa} l_f^{\kappa}) \times \kappa d^{1-\kappa} l_f^{\kappa-1} = \kappa(1 - \theta)1/l_f$. The above first-order condition implies that the solution for the woman's labor at home is

$$
l_f = \begin{cases}
\dfrac{(1 - \theta)\kappa[w_m + w_f - q]}{w_f[\theta + \kappa(1 - \theta)]}, & \text{for } d = \bar{\delta} \text{ (best);} \\[3mm]
\dfrac{(1 - \theta)\kappa[w_m + w_f]}{w_f[\theta + \kappa(1 - \theta)]}, & \text{for } d = \underline{\delta} \text{ (worst).}
\end{cases}
\tag{5.5}
$$

Note that the solution for l_f is a function of w_f, w_m, $p \in \{0, q\}$, and $d \in \{\underline{\delta}, \bar{\delta}\}$.

From the above two expressions, four facts can be deduced:

(1) The wife will work more in the market and less at home (so that l_f will be smaller) when the household purchases the bigger bundle of intermediate goods. This is due to the negative income effect associated with the necessity of paying the price $p = q > 0$ in order to buy the best technology, $d = \bar{\delta}$. The inferior technology is free, so that $p = 0$.

(2) Housework, l_f, is increasing in male's wage w_m. So, once again, a woman will work less in the market, if she has a well-off husband, other things being equal. This is due to the positive income effect associated with a rise in w_m.

(3) Suppose that the household purchases the superior technology. Housework, l_f, may fall or rise with an increase in the female's wage, w_f. With a little effort, it can be shown that

$$
\frac{dl_f}{dw_f} = -\frac{(1 - \theta)\kappa[w_m - q]}{w_f^2[\theta + \kappa(1 - \theta)]} \lessgtr 0, \text{ as } w_m - q \gtrless 0.
$$

There are income and substitution effects at work here. The substitution effect works to persuade the woman to work more in the market, when her wage rate goes up, and thus has a negative effect on l_f. The income effect has a positive impact on l_f. When a household gets more income, it would like to increase its consumption of both market and home goods. To increase its consumption of home goods, it must spend more time in household production, at least when the technology d is held fixed. One would expect a family in which the husband does not earn a lot to value an extra dollar of income more than one in which he does. Thus, the income effect of a rise in the woman's wage will be larger for a poorer family. This family will show a smaller decline (or even a rise) in l_f in response to an increase in w_f. Thus, one would expect dl_f/dw_f to be decreasing in $w_m - q$, and it is. When $w_m = q$, the income and substitution exactly cancel out. (In this situation the framework essentially collapses to the unisex single model because the man's earnings disappear after netting out q.) For a household that uses the inferior technology, housework will always drop with an increase in w_f; i.e., the substitution effect is always bigger than the income effect.

(4) A woman will spend more time in household production the more valuable her time is there, as captured by κ. This transpires because

$$\frac{dl_f}{d\kappa} = \frac{(1-\theta)w_f[w_m + w_f - p]\theta}{w_f^2[\theta + \kappa(1-\theta)]^2} > 0.$$

The term $w_m + w_f - p$ is greater than zero by the affordability condition; otherwise, the household would have negative consumption of nonmarket goods if they purchased the best household technology.

5.3   The Technology Adoption Decision, Stage 2

Which technology should the household pick, $d = \underline{\delta}$ or $d = \overline{\delta}$? This will depend on whether the maximal level of utility associated with adopting the worst technology, $V(w_f, w_m, 0, \underline{\delta})$, is bigger or smaller than the one with the best technology, $V(w_f, w_m, q, \overline{\delta})$. It is easy to deduce that

$$d = \begin{cases} \underline{\delta} \text{ (worst)}, & \text{if } V(w_f, w_m, 0, \underline{\delta}) > V(w_f, w_m, q, \overline{\delta}); \\ \overline{\delta} \text{ (best)}, & \text{if } V(w_f, w_m, 0, \underline{\delta}) \le V(w_f, w_m, q, \overline{\delta}). \end{cases}$$

Does there exist a unique threshold value for price of household inputs q, denoted by q^*, that solves the equation

$$V(w_f, w_m, 0, \underline{\delta}) = V(w_f, w_m, q^*, \overline{\delta})?$$

To answer this question, differentiate both sides of the maximization problem (5.3) with respect to d and p, separately. When doing this, a famous trick is employed.

5.3.1   The envelope theorem

When differentiating (5.3) with respect to p, you need to take into account that the optimal value of l_f will shift along with p. This is because l_f is an endogenous variable that is a function of p. This can be seen from (5.5). Doing the differentiation gives

$$\frac{dV(w_f, w_m, p, d)}{dp} = -\theta \frac{1}{w_m + w_f(1 - l_f) - p}$$

$$+ \underbrace{\left[-\theta \frac{w_f}{w_m + w_f(1 - l_f) - p} + (1 - \theta)\kappa \frac{1}{l_f}\right] \frac{dl_f}{dp}}_{= 0, \text{ by equation (5.4)}} < 0.$$

Observe that the term in brackets is equal to zero, a fact that follows from the first-order condition (5.4) for l_f. This is called the *envelope theorem*. In particular, it states that following a small change in some exogenous variable, the impact of a change in the choice variables on the value of the *optimized* objective function will wash out. Therefore,

$$\frac{dV(w_f, w_m, p, d)}{dp} = -\theta \frac{1}{w_m + w_f(1 - l_f) - p} < 0.$$

This expression can be interpreted easily. The left-hand side symbolically represents the shift in the household's maximal level of utility when p changes. The right-hand side shows how this is comprised. In particular, an increase in the price of household inputs by one unit will reduce the consumption of market goods by a unit. A one unit drop in market consumption leads to a fall in utility of $MU_c = \theta/c = \theta/[w_m + w_f(1 - l_f) - p]$.

Similarly,

$$\frac{dV(w_f, w_m, p, d)}{d(d)} = (1 - \theta)(1 - \kappa)\frac{1}{d} > 0.$$

A one unit increase in quantity of household inputs leads to a rise in the production of home goods by the marginal product of intermediate goods in home production, MP_d. It is straightforward to calculate from (5.1) that $MP_d = (1 - \kappa)d^{-\kappa}l_f^{\kappa}$. A one unit increase in home goods will cause utility to rise by $MU_n = (1 - \theta)/n = (1 - \theta)/(d^{1-\kappa}l_f^{\kappa})$, which is the marginal utility of home goods. Thus, the total marginal gain in utility is $MU_n \times MP_d = (1 - \theta)/(d^{1-\kappa}l_f^{\kappa}) \times (1 - \kappa)d^{-\kappa}l_f^{\kappa} = (1 - \theta)(1 - \kappa)d^{-1}$.

5.3.2   The choice of household technology: $d = \bar{\delta}$ (best) or $d = \underline{\delta}$ (worst)?

It will be shown that at a low price the household will pick $d = \bar{\delta}$, while at a high price it will choose $d = \underline{\delta}$. Now, as was just shown, $V(w_f, w_m, q, d)$ is decreasing in the price, q. Therefore, there exists a price, lying somewhere between the low and high one, at which the household is indifferent regarding the two technologies. Focus on the low price first. The above result implies

$$V(w_f, w_m, q, \bar{\delta}) > V(w_f, w_m, q, \underline{\delta}), \text{ for all } q,$$

because $\bar{\delta} > \underline{\delta}$ and $V(w_f, w_m, q, d)$ is increasing in d. Thus, by setting $q = 0$,

$$V(w_f, w_m, 0, \bar{\delta}) > V(w_f, w_m, 0, \underline{\delta}).$$

That is, if both household technologies are free, then the household would prefer the most productive one, or $d = \bar{\delta}$.

Next turn to the high price. Observe that as $q \to w_m + w_f$, the upper bound for the good technology to be affordable, $V(w_f, w_m, q, \bar{\delta}) \to -\infty$. This occurs because if the household buys the best technology at the price $q = w_m + w_f$, its consumption of market goods will be zero. And, $\lim_{c \to 0} \ln c = -\infty$. Hence,

$$V(w_f, w_m, w_m + w_f, \bar{\delta}) < V(w_f, w_m, 0, \underline{\delta}).$$

In other words, if the best household technology becomes prohibitively expensive, then the household will pick the less expensive, less productive one, or $d = \underline{\delta}$. So, at a very high price (say $q = w_m + w_f$) the household definitely prefers the $\underline{\delta}$ technology, while at a very low one (to wit $q = 0$) it will pick the $\bar{\delta}$ technology. Since $V(w_f, w_m, q, \bar{\delta})$ is decreasing in q there will exist a single intermediate price, q^*, that results in $V(w_f, w_m, q^*, \bar{\delta}) = V(w_f, w_m, 0, \underline{\delta})$.

To conclude, there exists a unique threshold value for q, denoted by q^*, such that

$$V(w_f, w_m, q^*, \bar{\delta}) = V(w_f, w_m, 0, \underline{\delta}).$$

This situation is portrayed in figure 1.12. The threshold price is given by q^*, the point where the $V(w_f, w_m, q, \bar{\delta})$ and $V(w_f, w_m, 0, \underline{\delta})$ curves intersect. The $V(w_f, w_m, q, \bar{\delta})$ curve slopes downward because $dV(w_f, w_m, q, \bar{\delta})/dq < 0$, as demonstrated above. The $V(w_f, w_m, 0, \underline{\delta})$ curve is simply some constant, as it is not a function of q. Note that the $V(w_f, w_m, q, \bar{\delta})$ curve starts above the $V(w_f, w_m, 0, \underline{\delta})$ curve, when $q = 0$, and ends below when $q = w_m + w_f$. Therefore, since $V(w_f, w_m, q, \bar{\delta})$ is continuously decreasing in q, the two curves must intersect and do so just once. At any price $q > q^*$, the $V(w_f, w_m, q, \bar{\delta})$ curve lies below the $V(w_f, w_m, 0, \underline{\delta})$ one. Thus, the household will not pick $\bar{\delta}$. When $q < q^*$, the $V(w_f, w_m, q, \bar{\delta})$ curve lies above the $V(w_f, w_m, 0, \underline{\delta})$ one, the household will adopt $\bar{\delta}$.

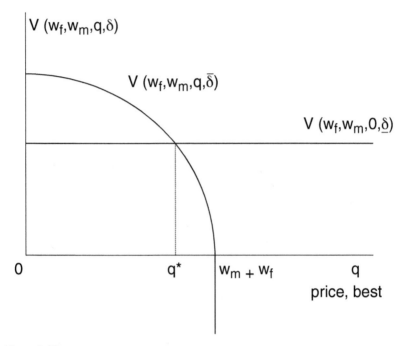

Figure 1.12
Household technology adoption decision.

5.4 The Rise in Married Female Labor Supply—A Prelude

Suppose that the household technology is now represented by the triplet (d, p, κ) that can can be either $(\underline{\delta}, 0, \underline{\kappa})$ or $(\bar{\delta}, q, \bar{\kappa})$, with $\bar{\kappa} < \underline{\kappa}$, corresponding to the worst and best technologies. That is, suppose that the two household technologies are now given by $n = \underline{\delta}^{1-\kappa} l_f^{\underline{\kappa}}$, with a price of 0, and $n = \bar{\delta}^{1-\bar{\kappa}} l_f^{\bar{\kappa}}$, with a price of $q > 0$, where $\bar{\kappa} < \underline{\kappa}$. So, in addition to providing a greater quantity of the intermediate goods used in household production, the best technology also places less value (a lower κ) on the labor used at home. (The discussion presupposes that the best technology is preferred to the worst one when the former is free.) Imagine a situation where the price of the intermediate good used in household production is declining over time. Households may differ in their threshold prices, perhaps because of differences in wages across households or in the values that they place on household goods. When the price is high, very few households will purchase it. As the price crosses the threshold for a household, it will purchase the good. This will be associated with a rise in married female labor supply for two reasons. First, the advanced technology costs more. There will be a negative income effect associated with the adoption of the advanced technology. This operates to increase female labor supply at the time of adoption. Second, the advanced technology relies less on a woman's time in household production, so it sheds household labor.

6 The Rise in Married Female Labor Supply

Is it, then, consistent to hold the developed woman of this day within the same narrow political limits as the dame with the spinning wheel and knitting needle occupied in the past? No, no! Machinery has taken the labors of woman as well as man on its tireless shoulders; the loom and the spinning wheel are but dreams of the past; the pen, the brush, the easel, the chisel, have taken their places, while the hopes and ambitions of women are essentially changed.

—Elizabeth Cady Stanton, "Solitude of Self," address before United States Congressional Committee on the Judiciary, Monday January 18, 1892

For ages woman was man's chattel, and in such condition progress for her was impossible; now she is emerging into real sex independence, and the resulting outlook is a dazzling one. This must be credited very largely to progression in mechanics; more especially to progression in electrical mechanics.

Under these new influences woman's brain will change and achieve new capabilities, both of effort and accomplishment.

—Thomas Alva Edison, as interviewed in *Good Housekeeping Magazine* 55, no. 4 (1912): 440.

6.1 Some Economic History

In 1900 only 5 percent of married women worked. This had risen to 61 percent by 2000. The trend in married female labor-force participation is displayed in the right-hand panel of figure 1.13. What can explain this rise in female labor supply? A clue is provided in the left-hand-side panel of the figure. As female labor supply rose, the amount of time spent on housework declined. In 1900 the average household spent 58 hours a week on housework, including meal preparation, laundry and cleaning.[7] This figure was just 18 in 1975. At the same time, the number of paid domestic workers declined, presumably due in part to the labor-saving nature of household appliances. Hence, the time spent on the more onerous household chores, such as those associated with cooking, cleaning, doing laundry, and so on, declined considerably over the last century. Other things also

---

7. Reid (1934, table 21) reports similar numbers from a study of 442 Oregon families. In 1900 the American economy was very rural.

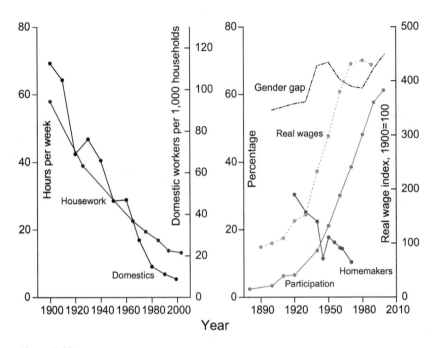

Figure 1.13
Housework and market work.
Source: Greenwood, Seshadri, and Yorukoglu 2005 (partially updated).

changed, including a four-fold rise in the general level of real wages. Real wages rose due to technological progress in the market sector that increased the marginal product of labor. That is, for any given level of employment in the market sector, an extra unit of labor could produce more over time. This made labor more valuable. As was discussed earlier, with logarithmic preferences of the sort that macroeconomists typically use, this rise in the general level of wages would have no effect on married female labor supply.

The gender wage gap also shrunk. In 1900 a working woman earned about 50 percent of what a man did, and by 2000 this number had risen to 72 percent. (Statistics concerning the wage gap are discussed further in section 7.) The earlier analysis suggested that theoretically speaking this shrinkage could lead to a rise in married female labor supply. Whether or not this rise in the relative earnings of a working woman can explain on its own the dramatic rise in married female labor-force participation will depend on the size of the elasticity of married female labor supply. On this, a caveat is in order. Taking the gender wage gap as an exogenous variable is a bit precarious. As will be discussed, it may be a function of technological considerations in the economy. For example, mechanization has resulted in mental skills becoming more valuable relative to physical strength. This makes the job market more attractive to women, as will be discussed in section 8. Additionally, the decline in fertility, improvements in household technologies, and advances in obstetric and pediatric medicine have reduced the time off of work that a woman needs to bear and raise children. This influences the types of jobs that she will take. (This factor is analyzed in section 6 of chapter 2.) Also, as will be seen, even the extent of discrimination against married women in the workplace, which influences the gender gap, may be a function of the economic environment. (This is covered in section 1 of chapter 4.)

The Second Industrial Revolution occurred at the beginning of the twentieth century. This era is connected with the introduction of electricity, the automobile, and the petrochemical industry. But it also saw the inception of central heating, dryers, electric irons, frozen foods, refrigerators, sewing machines, washing machines, vacuum cleaners, and other appliances now considered fixtures of everyday life. The spread through the U.S. economy of electricity, central heating, flush toilets, and running water is shown in figure 1.14, which plots the percentage of households that adopted the basic facilities in question.

Figure 1.14 also plots the diffusion through American households of some common electrical appliances, which was spurred on by a rapid drop in their prices. Figure 1.15 shows the fall in the quality-adjusted time price for select appliances in the postwar period. (The time price measures the amount of time that an average person would need to work in order to purchase the appliance. This price is presented in an index form.) Because appliances have changed drastically over time, economists attempt to control for the change in the quality of these goods when they measure their prices. Figure 1.16 shows what a washing machine looked like in 1911. The contraption on the right-hand side of the picture was used for churning butter. The impact of appliances on the lives of Americans was gradual. It took time for both their prices to fall and their quality to improve.

To understand the impact that the Second Industrial Revolution had on the home, try to imagine the tyranny of household chores at the turn of the last century. In 1890 only 24 percent of houses had running water,

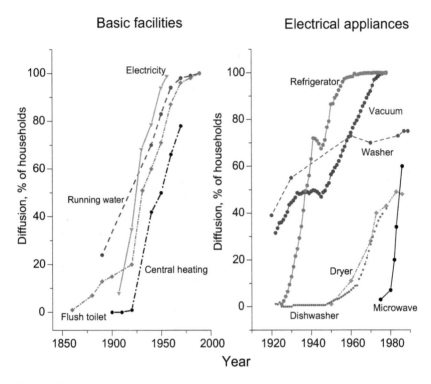

Figure 1.14
The diffusion of basic facilities and electrical appliances through the U.S. economy.
Source: Greenwood, Seshadri, and Yorukoglu 2005.

and none had central heating. So each year the average household lugged
7 tons of coal and 9,000 gallons of water around the home. The simple task
of laundry was a major operation in those days. While mechanical washing machines were available as early as 1869, this invention really took off
only with the development of the electric motor. Ninety-eight percent of
households still used a 12-cent scrubboard to wash their clothes in 1900.
Water had to be transported to the stove, where it was heated by burning
wood or coal. Once the clothes were cleaned via a washboard or mechanical washing machine, they had to be rinsed and then be wrung out either
by hand or by using a mechanical wringer. After this, the clothes were
hung out to dry on a clothes line. Then, the oppressive task of ironing
began, using heavy flatirons that had to be heated continuously on the
stove.

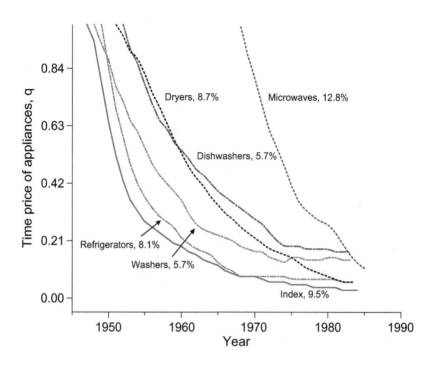

Figure 1.15
Quality-adjusted time prices for select appliances.
Source: Greenwood, Seshadri, and Yorukoglu 2005.

The amount of time freed by modern appliances is somewhat specula-
tive. Controlled engineering studies documenting the time saved on some
specific task by the use of a particular machine would be ideal. Unfortu-
nately, these studies are hard to come by. In one case, though, the Rural
Electrification Authority supervised one such study based on 12 farm wives
during 1945–1946. They compared the time spent doing laundry by hand
to that spent using electrical equipment. The upshot of the study is shown
in table 1.3. The women also wore a pedometer. One subject, Mrs. Verett,
was reported on in detail.[8] Without electrification, she did her laundry
in the manner described above. (She actually used a gas-powered washing
machine instead of a scrubboard.) After electrification Mrs. Verett had an
electric washer, dryer, and iron. A water system was also installed with a

8. This study is reported in *Electrical Mechandising*, March 1, 1947, 13 and 38–39.

Figure 1.16
A 1911 Maytag washing machine.

water heater. The researchers estimated that it took her about 4 hours to do
a 38-pound load of laundry by hand, and then about 4.5 hours to iron it
using old-fashioned irons. By comparison it took 41 minutes to do a load
of laundry using electrical appliances and 1.75 hours to iron it. The woman
walked 3,181 feet to do the laundry by hand, and only 332 feet with electrical equipment. She walked 3,122 feet when ironing the old way, and 333
the new way.

At the start of the Second Industrial Revolution women's magazines were
filled with articles extolling the virtues of appliances, the new domestic
servants. For example, in 1920 an article in the *Ladies' Home Journal* entitled
"Making Housekeeping Automatic" claimed that appliances could save a
4-person family 18.5 hours a week in housework (see table 1.4).

While the development of new consumer durables was important
in liberating women from the shackles of housework, so too was the
rationalization of the household. The principles of scientific management

Table 1.3
Time spent washing and drying a load of laundry.

	Without appliances		With appliances	
Task	Time	Distance	Time	Distance
Washing	4 hrs	3,181 ft	41 min	332 ft
Ironing	3.5 hrs	3,122 ft	1.75 hrs	333 ft

Source: R. Gaffney, "The Complete FARM." *Electrical Merchandising*, (March 1, 1947), 13 and 38–39.

Table 1.4
Estimated weekly hours saved by appliances.

Task	With appliances	Without appliances	Time savings
Breakfast	7	10	3
Luncheons	10.5	14	3.5
Dinners	10	12	2
Dishwashing and clearing	10.5	15.75	5.25
Washing and ironing	6.5	9	2.5
Marketing and errands	6	6	0
Sewing and mending	3.5	4	0.5
Bed making	2.75	3.5	0.75
Cleaning and dusting	2	3	1
Cleaning kitchen and refrigerator	2	2	0
Total	60.75	79.25	18.5

Source: R. McMahon. "Making Housekeeping Automatic." *Ladies' Home Journal* 37 (September 1920), 3–4.

were applied to the home, just as in the factory. Domestic tasks were studied with the aim of improving their efficiency. Christine Frederick was an early advocate of applying the principals of scientific management to the home. She was captivated by the fact that a man named Frank B. Gilbreth had been able to increase the output of bricklayers from 120 to 350 bricks per hour by applying the principles of scientific management. He did this by placing an adjustable table by the bricklayer's side so that the latter wouldn't have to stoop down to pick up a brick. He also had the bricks delivered on it in the right position so that there would be no need for the bricklayer to turn each one right-side up. Additionally, he taught bricklayers to pick bricks up with their left hands and simultaneously take trowelfuls of mortar with their right hands.

Frederick (1912) applied the idea to dishwashing first and then to other tasks. She broke dishwashing down into three separate tasks: scraping and stacking; washing; and drying and putting the dishes and utensils away. She computed the correct height for sinks. She discovered that dishwashing could be accomplished more efficiently by placing drainboards on the left, using deeper sinks, and by connecting a rinsing hose to the hot water outlet; she estimated that this saved 15 minutes per dinner. In 1913 she wrote: "Didn't I with hundreds of women stoop unnecessarily over kitchen tables, sinks, and ironing boards, as well as bricklayers stoop over bricks?"[9] Frederick and others in the home economics movement had a tremendous impact on the design of appliances and houses. Take the kitchen, for example. The kitchen of the 1800s was characterized by a large table and isolated cupboard. An organized kitchen with continuous working surfaces and built-in cabinets began to appear in the 1930s, after a period of slow evolution. The kitchen became connected with the dining room and other living areas in the 1940s, ending the housewife's isolation. Figure 1.17 pictures what a kitchen on a farm would have looked like in 1900. Figure 1.18 shows two diagrams by Christine Frederick illustrating a poorly (left) and efficiently (right) organized kitchen. In the efficiently organized kitchen, all the equipment used either to prepare (A) or clear (B) food is arranged in close proximity. In the poorly planned kitchen the arrangement of equipment is more haphazard.

In 1924 a pair of famous sociologists, Robert and Helen Lynd, studied Middletown, a small town in Indiana. They found that 87 percent of married women spent 4 or more hours doing housework each day. Zero percent spent less than 1 hour a day. The town was restudied by sociologists at two later dates. Table 1.5 displays the findings. By 1999 only 14 percent of married women spent more than 4 hours a day on housework, and 33 percent spent less than 1 hour a day. Definitive recent evidence on the decline in time spent on housework is provided by the Aguiar and Hurst (2007) study discussed earlier.

9. As quoted in Giedion (1948, 521).

Figure 1.17
A kitchen on a farm in 1900.

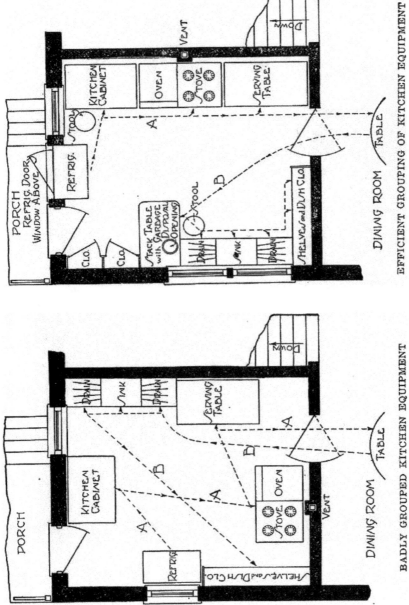

BADLY GROUPED KITCHEN EQUIPMENT

EFFICIENT GROUPING OF KITCHEN EQUIPMENT

Figure 1.18

"A" shows the preparing route, while "B" denotes the clearing one.

Source: Frederick (1919, 22–23) *Household Engineering.*

Table 1.5
Daily housework in Middletown, IN.

	Percentage of married housewives in each category		
Year	≥ 4 hours	2 to 3 hours	≤ 1 hour
1924	87	13	0
1977	43	45	12
1999	14	53	33

Source: Caplow, Hicks, and Wattenberg 2001, 37.

To conclude this section, poorer countries today may partially reflect the United States in the past. While a fraction of citizens of poorer countries may use modern appliances, the rest won't utilize them because of their high cost. The relationship between time spent on either cleaning or cooking, on the one hand, and per-capita GDP, on the other. is plotted in figure 1.19. As can be seen, average weekly hours spent on cleaning or cooking and per-capita GDP is negatively correlated. The Pearson correlation coefficients are -0.48 and -0.76, respectively. Time spent on core household chores, such as cleaning and cooking, has declined with economic development.

6.2 Household Production Theory, Again

Household production theory is used now to examine the rise in married female labor supply. Again let the household's tastes be represented by

$$u = \theta \ln c + (1 - \theta) \ln n, \text{ with } 0 < \theta < 1,$$

where c gives the consumption of market goods and n denotes the consumption of home ones. Assume that the household produces home goods, n, according to the following *constant-elasticity-of-substitution* (CES) production function

$$n = [\kappa \delta^\rho + (1 - \kappa) l_f^\rho]^{1/\rho}, \text{ for } 0 < \kappa < 1 \text{ and } \rho \leq 1,$$

where again δ is the amount of intermediate goods used in household production and l_f stands for the amount of the wife's household labor. The parameter ρ controls the degree of substitution between intermediate goods and labor in household production. Observe that when $\rho = 1$, the above production function is linear. In this situation intermediate goods and labor

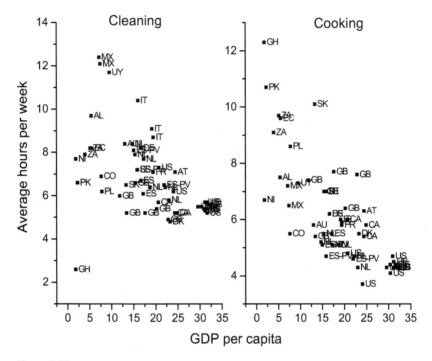

Figure 1.19
The cross-country relationship between average hours on housework per week and
GDP per capita. A nation is labeled by its standard two letter country code. Some
countries have more than one data point, corresponding to different years and hence
different levels of income. Per-capita income is reported in terms of 1,000s of 1990$.
Source: Bridgman, Duernecker, and Herrendorf 2017.

are highly substitutable. When $\rho = 0$, the production function approaches
the Cobb-Douglas form given by (5.1). Last, when $\rho = -\infty$ the production
function takes the *Leontief* form $n = \min\{\kappa\delta, (1-\kappa)l\}$.[10] In this case there
is no substitution across inputs. The *elasticity of substitution* between intermediate goods and labor is given by $1/(1-\rho)$. The parameter ρ plays an
important role in the analysis. Intermediate goods can be purchased at the
time price q. This measures the amount of time working that it would take
for a woman to purchase these goods. Intermediate goods are assumed to
be purchased in the fixed amount δ.

10. The last two facts are demonstrated in Varian (1978).

Definition 2 (Elasticity of Substitution) *Suppose that intermediate goods and labor, δ and l, are used in production. Let the goods price of these inputs be $\widetilde{q}$ and w, respectively. The elasticity of substitution gives the percentage change in the input ratio, δ/l, in response to some percentage change in the price ratio, $\widetilde{q}/w \equiv q$. It measures how a change in the relative price of inputs affects the relative use of inputs. Specifically, it is defined as*

$$-\frac{\widetilde{q}/w}{\delta/l}\frac{d(\delta/l)}{d(\widetilde{q}/w)} = -\frac{d\ln(\delta/l)}{d\ln(\widetilde{q}/w)} = -\frac{d\ln(\delta/l)}{d\ln(q)}.$$

Note that since a rise in $\widetilde{q}/w$ should lead to a decline in δ/l, the elasticity is usually normalized to be a positive number, which explains the minus sign. The term $q \equiv \widetilde{q}/w$ is the relative price of intermediate goods measured in units of time—i.e. the time price of intermediate goods.

Example 7 (CES production function) *Imagine a firm that hires intermediate goods and labor, δ and l, at the prices $\widetilde{q}$ and w, in order to produce output, as given by $[\kappa\delta^\rho + (1-\kappa)l^\rho]^{1/\rho}$. The firm seeks to maximize profits. Its maximization problem is*

$$\max_{\delta,l}\{[\kappa\delta^\rho + (1-\kappa)l^\rho]^{1/\rho} - \widetilde{q}\delta - wl\}.$$

The first-order conditions for δ and l are, respectively:

$$[\kappa\delta^\rho + (1-\kappa)l^\rho]^{1/\rho-1}\kappa\delta^{\rho-1} = \widetilde{q},$$

and

$$[\kappa\delta^\rho + (1-\kappa)l^\rho]^{1/\rho-1}(1-\kappa)l^{\rho-1} = w.$$

Divide the first condition by the second to get

$$\frac{\kappa\delta^{\rho-1}}{(1-\kappa)l^{\rho-1}} = \frac{\widetilde{q}}{w} = q.$$

Again, in the above expression, $q \equiv \widetilde{q}/w$ is the time price of intermediate goods. The equation can be expressed in logs as

$$\ln[\kappa/(1-\kappa)] + (\rho-1)\ln(\delta/l) = \ln(\widetilde{q}/w).$$

Therefore,

$$-\frac{d\ln(\delta/l)}{d\ln(\widetilde{q}/w)} = -\frac{d\ln(\delta/l)}{d\ln(q)} = -\frac{1}{\rho-1} = \frac{1}{1-\rho},$$

so that the elasticity of substitution is increasing in ρ. Observe that when (i) $\rho = 1$ (the linear case) the elasticity of substitution is infinite; (ii) when $\rho = 0$ (the

Cobb-Douglas case) the elasticity of substitution is 1; and (iii) when $\rho = -\infty$ (the Leontief case) the elasticity of substitution is 0.

6.3 The Couple's Maximization Problem

The household's decision problem is easily seen to be

$$\max_{l_f}\{\theta \ln[w_m + w_f(1 - l_f) - w_f q] + (1 - \theta)(1/\rho) \ln[\kappa \delta^\rho + (1 - \kappa)l_f^\rho]\}.$$

The first-order condition linked to this problem is

$$\theta \underbrace{\frac{1}{w_m + w_f(1 - l_f) - w_f q}}_{MU_c} \times w_f = (1 - \theta) \underbrace{\frac{(1 - \kappa)l_f^{\rho-1}}{\kappa \delta^\rho + (1 - \kappa)l_f^\rho}}_{MU_{l_f} = MU_n \times MP_{l_f}}.$$

The left-hand side represents the marginal cost, MC, of spending an extra unit of time at home. By doing so, the family loses w_f in wages or market consumption. A unit drop in market consumption will lead to a loss of $MU_c = \theta/c = \theta/[w_m + w_f(1 - l_f) - w_f q]$ in utility terms. The right-hand side gives the marginal benefit, MB, from working an extra unit at home. Think about $(1 - \theta)(1/\rho) \ln[\kappa \delta^\rho + (1 - \kappa)l_f^\rho]$ as representing the utility function for l_f. Then, an extra unit of time spent at home will cause utility to rise by $MU_{l_f} = (1 - \theta)(1 - \kappa)l_f^{\rho-1}/[\kappa \delta^\rho + (1 - \kappa)l_f^\rho]$. Now, the marginal utility of the woman's time spent at home, MU_{l_f}, can be further decomposed into the product of two terms. An extra unit of home work, l_f, will raise the production of home goods by the marginal product of the woman's time, $MP_{l_f} = [\kappa \delta^\rho + (1 - \kappa)l_f^\rho]^{1/\rho-1}(1 - \kappa)l_f^{\rho-1}$. An extra unit of home goods leads to gain in utility of $MU_n = (1 - \theta)/n = (1 - \theta)/[\kappa \delta^\rho + (1 - \kappa)l_f^\rho]^{1/\rho}$. Therefore, the marginal benefit of spending an extra unit of time at home is $MP_{l_f} \times MU_n = [\kappa \delta^\rho + (1 - \kappa)l_f^\rho]^{1/\rho-1}(1 - \kappa)l_f^{\rho-1} \times (1 - \theta)/[\kappa \delta^\rho + (1 - \kappa)l_f^\rho]^{1/\rho} = (1 - \theta)(1 - \kappa)l_f^{\rho-1}/[\kappa \delta^\rho + (1 - \kappa)l_f^\rho] = MU_{l_f}$.

The above first-order condition can be rewritten as

$$\theta \underbrace{\frac{1}{(1/\phi) + (1 - l_f) - q}}_{MC = LHS} = (1 - \theta) \underbrace{\frac{(1 - \kappa)}{\kappa \delta^\rho l_f^{1-\rho} + (1 - \kappa)l_f}}_{MB = RHS}. \tag{6.1}$$

In the above expression $\phi \equiv w_f/w_m$, which represents the gender wage gap. The situation is displayed in figure 1.20, which plots the left- and right-hand sides of (6.1) as a function of amount of time the woman spends at

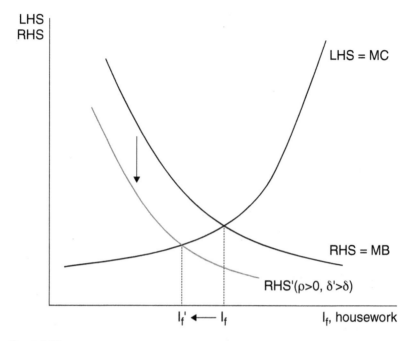

Figure 1.20
Determination of housework, l_f. The figure also illustrates the impact on housework, l_f, of an increase in household capital from δ to δ', when $\rho > 0$.

home, l_f. The left-hand side is easily seen to be rising in l_f. This is because the more time a woman spends at home, the more valuable an extra unit of market consumption becomes due to the diminishing marginal utility of market goods. The right-hand side is decreasing in l_f, since $1 - \rho > 0$. This transpires because as more time is devoted to the home, the less valuable an extra unit of home goods becomes due both to diminishing returns in household production and the diminishing marginal utility of home goods.

6.4 Graphical Analysis of the Rise in Married Female Supply

How will an increase in the stock of household capital, δ, affect hours worked by a woman at home, l_f? The left-hand side of (6.1) is not a function of δ. Hence, it will not change. Focus on the denominator of the right-hand side. In the denominator is the term δ^ρ. This is increasing or decreasing in δ depending on whether $\rho \gtreqless 0$. Therefore, the right-hand side will increase or decrease as $\rho \lesseqgtr 0$. More formally, take the derivative of the right-hand side of (6.1) with respect to δ. One gets

$$\frac{d\text{RHS}}{d\delta} = -\rho(1-\theta)\frac{(1-\kappa)\kappa\delta^{\rho-1}l_f^{\rho-1}}{[\kappa\delta^\rho + (1-\kappa)l_f^\rho]^2} \gtreqless 0, \text{ as } \rho \lesseqgtr 0.$$

Therefore, the right-hand side curve shifts down (up) when $\rho > 0$ (< 0). Hence, hours worked at home, l_f, fall when $\rho > 0$ and rise when $\rho < 0$. Thus, hours worked in the market, $1 - l_f$, rise when capital and household labor are more substitutable than they would be with a Cobb-Douglas production function ($\rho > 0$) and fall when they are less substitutable ($\rho < 0$).

Why should the response of a woman's work at home to a rise in the stock of household capital, δ, depend upon ρ? To understand this, consider the second term in the household's objective function. From this it can be seen that the marginal utility of housework is

$$\text{MU}_{l_f} = \frac{d\{(1-\theta)(1/\rho)\ln[\kappa\delta^\rho + (1-\kappa)l_f^\rho]\}}{dl_f} = (1-\theta)\frac{(1-\kappa)l_f^{\rho-1}}{\kappa\delta^\rho + (1-\kappa)l_f^\rho},$$

which is the right-hand side of the first-order condition (6.1). Take the derivative of the above expression with respect to δ to get

$$\frac{d\text{MU}_{l_f}}{d\delta} = \frac{d^2\{(1-\theta)(1/\rho)\ln[\kappa\delta^\rho + (1-\kappa)l_f^\rho]\}}{dl_f d\delta}$$

$$= -\rho(1-\theta)\frac{(1-\kappa)\kappa\delta^{\rho-1}l_f^{\rho-1}}{[\kappa\delta^\rho + (1-\kappa)l_f^\rho]^2} = \frac{d\text{RHS}}{d\delta}.$$

Now, when $\rho > 0$, an increase in δ reduces the marginal utility of housework. Therefore, less will be done. When $\rho < 0$, an increase in δ raises the marginal utility of housework. Thus, more will be done. When $\rho > 0$, an increase in δ will cause the right-hand side to shift down. Thus, home work, l_f, falls. The opposite obtains when $\rho < 0$. This relates to the notion of Edgeworth-Pareto substitutes and complements. When $\rho < 0$, household capital, δ, and home hours, l_f, are Edgeworth-Pareto complements in production, while if $\rho > 0$, household capital and housework are Edgeworth-Pareto substitutes.

Definition 3 (Edgeworth-Pareto Substitutes and Complements) *Let $U(x, y)$ be a utility function in two goods x and y. Then, $U_1(x, y) = dU(x, y)/dx$ is the marginal utility of x and $U_2(x, y) = dU(x, y)/dy$ is the marginal utility of y. If $U_{12}(x, y) = d^2U(x, y)/dxdy < 0$, the goods are called Edgeworth-Pareto substitutes,*

while if $U_{12}(x, y) = d^2 U(x, y)/dxdy > 0$, *they are Edgeworth-Pareto complements.*
Note that $U_{12}(x, y) = U_{21}(x, y)$, *by Young's theorem. Hence, if* x *and* y *are*
Edgeworth-Pareto substitutes, more of one good reduces the marginal utility of
the other, while if x *and* y *are Edgeworth-Pareto complements, more of one good*
raises the marginal utility of the other.

In a similar vein, observe that a decrease in $1/\phi$ (or a reduction in
the gender wage gap) will cause the left-hand side of (6.1) to shift up.
The right-hand side remains constant. Homework, l_f, will fall as a result.
Equiproportional increases in the female and male wages, w_f and w_m, have
no impact on housework; all that matters are shifts in the gender wage gap.

To conclude this section:

(1) Technological progress in the household sector in the form of new and
improved intermediate inputs, as measured by an increase in δ, will cause
housework, l_f, to fall when intermediate goods and labor are Edgeworth-
Pareto substitutes ($\rho > 0$) in home production.

(2) An equal (proportionate) increase in women's and men's wages, w_f and
w_m, will have no impact on married female labor supply, $1 - l_f$.

(3) A narrowing of the gender wage gap, as reflected by a rise in ϕ, will lead
to more married women working in the market.

6.5 Mathematical Analysis of the Rise in Married Female Supply

Equation (6.1) does not have a closed-form solution for l_f. So, how can it
be analyzed?[11] Equation (6.1) can be rewritten as

$$\kappa \delta^\rho + (1 - \kappa)l_f^\rho = \frac{(1 - \theta)}{\theta}(1 - \kappa)[(1/\phi + 1 - q)l_f^{\rho-1} - l_f^\rho]. \tag{6.2}$$

Suppose one wants to examine the impact of δ on l_f. Then, take the total
differential of the above equation with respect to δ and l_f:

$$\rho \kappa \delta^{\rho-1} d\delta + \rho(1 - \kappa)l_f^{\rho-1} dl_f = (\rho - 1)\frac{(1 - \theta)}{\theta}(1 - \kappa)l^{\rho-2}(1/\phi + 1 - q)dl_f$$

$$- \frac{(1 - \theta)}{\theta}\rho(1 - \kappa)l_f^{\rho-1} dl_f.$$

(Total differentials are discussed in the Mathematical Appendix.) Using the
fact from (6.2) that $[(1 - \theta)/\theta](1 - \kappa) = [\kappa \delta^\rho + (1 - \kappa)l_f^\rho]/[(1/\phi + 1 - q)l_f^{\rho-1} - l_f^\rho]$

11. Those not interested in a mathematical derivation of the rise in married female
labor-force participation can omit this subsection.

in the above equation gives

$$[(1/\phi + 1 - q)l_f^{\rho-1} - l_f^{\rho}]\rho\kappa\delta^{\rho-1}d\delta$$

$$+ \rho(1 - \kappa)l_f^{\rho-1}[(1/\phi + 1 - q)l_f^{\rho-1} - l_f^{\rho}]dl_f$$

$$= (\rho - 1)l_f^{\rho-2}(1/\phi + 1 - q)[\kappa\delta^{\rho} + (1 - \kappa)l_f^{\rho}]dl_f$$

$$- \rho l_f^{\rho-1}[\kappa\delta^{\rho} + (1 - \kappa)l_f^{\rho}]dl_f.$$

With some effort, it can be shown that

$$\frac{dl_f}{d\delta} = -\rho\frac{[(1/\phi + 1 - q) - l_f]l_f\kappa\delta^{\rho-1}}{(1/\phi + 1 - q)[\kappa\delta^{\rho} + (1 - \kappa)l_f^{\rho}] - \rho[(1/\phi + 1 - q - l_f]\kappa\delta^{\rho}} \gtreqless 0, \text{ as } \rho \lesseqgtr 0.^{12}$$

Likewise, it can be deduced that

$$\frac{dl_f}{d(1/\phi)} = \frac{(1 - \kappa)(1 - \theta)/\theta}{(1/\phi + 1 - q)[\kappa\delta^{\rho} + (1 - \kappa)l_f^{\rho}] - \rho[(1/\phi + 1 - q - l_f]\kappa\delta^{\rho}} > 0.$$

7   The Gender Wage Gap

Female workers have earned less than male workers in the United States, on average, as the right-hand-side panel of figure 1.13 shows. This is not just a U.S. phenomenon. There is a gender wage gap in all countries of the

---

12. The calculations are continued from here. Move the dl_f term on the left onto the right-hand side. Expanding some terms allows for some cancellations, as can be seen from the series of calculations below.

$$[1/\phi + 1 - q - l_f]l_f^{\rho-1}\rho\kappa\delta^{\rho-1}d\delta = \rho(1 - \kappa)l_f^{2\rho-1}dl_f$$

$$- l_f^{\rho-2}(1/\phi + 1 - q)[\kappa\delta^{\rho} + (1 - \kappa)l_f^{\rho}]dl_f$$

$$+ \rho l_f^{\rho-2}(1/\phi + 1 - q)\kappa\delta^{\rho}dl_f$$

$$- \rho l_f^{\rho-1}[\kappa\delta^{\rho} + (1 - \kappa)l_f^{\rho}]dl_f$$

$$= -l_f^{\rho-2}(1/\phi + 1 - q)[\kappa\delta^{\rho} + (1 - \kappa)l_f^{\rho}]dl_f$$

$$+ \rho l_f^{\rho-2}[1/\phi + 1 - q - l_f]dl_f.$$

Now, divide both sides by $l_f^{\rho-1}$ and rearrange. The desired end result then obtains. Note that the denominator in the above expression is positive because $\rho \leq 1$ and $1/\phi + 1 - q - l_f < 1/\phi + 1 - q$. The graphical analysis is easier!

Organisation for Economic Co-operation and Development (OECD). So, for example, in 2000 the weekly or monthly earnings of women was 83 percent of that of men in Australia, 80 percent in West Germany, 65 percent in Japan, 82 percent in New Zealand, and 76 percent in the United Kingdom, compared with 75 percent in the United States. If one looks at annual earnings instead, the gender wage gap was 80 percent in Finland, 91 percent in France, 78 percent in the Netherlands, and 85 percent in Sweden. Understanding the gender wage gap is complicated and nuanced, unlike the simplistic analyses that politicians and reporters offer. The gender wage gap is due to many factors, not just discrimination. It may reflect differences in educational attainment and occupational choice. There is also a tendency for women to experience career interruptions, to take part-time jobs, or to refuse overtime so that they can free up time for raising children, another form of work. Also, women may prefer to take less home in wages in exchange for better benefits, such as child care, parental leave, and sick leave. As will be discussed now, adjusting for such factors significantly reduces the measured gender wage gap.

Three measures of the gender wage gap are shown for 1980 and 2010 in figure 1.21. Focus on the "raw" measure first. This is essentially the ratio of the average wage earned by a woman relative to that of a man. As can be seen, it narrowed considerably between 1980 and 2010. Most of the shrinkage in the gender wage gap occurred in the 1980s. Still, despite the narrowing of the raw measure, in 2010 women earned roughly only 80 percent of what men did. Part of the gender wage gap is accounted for by the fact that women have different levels of education and job experience than men. The second measure statistically adjusts for this (as well as for other things such as race, region, and metropolitan area). Differences in education and work experience between men and women accounted for a significant amount of the gender wage gap in 1980, as can be seen by the difference between the first and second measures. In 2010, however, the gap between the first and second measures was small, because women increased their human capital relative to men, both in terms of education and experience. In 2010 women had higher levels of educational attainment than men, but less job experience. Men and women may choose different occupations, work in different industries, and have different degrees of union coverage. (Unions tend to equalize wages across individuals.) For example,

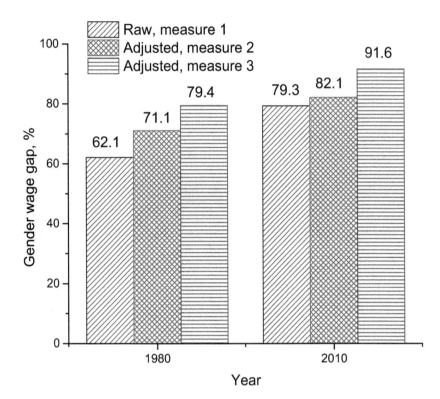

Figure 1.21
Gender wage gaps in 1980 and 2010, three measures.
Source: Blau and Kahn 2017, figure 2.

in 2011 over 80 percent of elementary and kindergarten teachers, librarians, and nurses were women. But, roughly over 80 percent of architects, computer programmers, dentists, and engineers were men (this number was higher still in 1970). The third measure adjusts for these factors as well. Such differences accounted for a large part of the gender wage gap in 1980 and still did in 2010, as reflected by the difference between the second and third measures. Note that in 2010 there still remained a gap between men's and women's wages of about 8 percent.

Does this imply that discrimination resulted in women earning 8 percent less than men? It is difficult to answer this question. On the one hand, suppose discrimination does exist. Then, it will influence the occupations and industries that women choose to work in. So, then 8 percent would be an underestimate. Historically speaking, discrimination against married women in the workplace must have affected the education that they acquired: Why undertake advanced schooling if you can't work? On the other hand, the gap may reflect characteristics that are unobservable to the statistician. For example, married women with children may prefer jobs that have better benefits, regular hours, and less risk of unemployment, while paying less in terms of wages. The above measures do not control for a woman's marital status and the number of children she has, which is a shortcoming. Some research suggests that the gender wage gap for unmarried childless women is quite small.

Three models of the gender wage gap will be presented. Two models highlight the fact that the gender wage gap may be a function of economic considerations, such as the type of jobs that were available historically for women (see section 8 in this chapter) or how the necessity to take time off of work to bear and raise children influences the type of jobs that women take (section 6 in chapter 2). The third (section 1 in chapter 4) focuses on discrimination in the labor market. It models how discrimination against married women in the workforce changed over time as a function of the economic environment; i.e., even discrimination may have an endogenous component to it.

8 Brain and Brawn

Over time physical labor has become progressively less important relative to mental labor. The right-hand-side panel of figure 1.22 illustrates the fraction of all non-farm jobs made up by blue- and white-collar jobs. Blue-collar occupations consist of craft workers, service workers, laborers, and operatives. White-collar occupations include clerical workers, managers, professionals, proprietors, and sales workers. In 1860 the vast majority of jobs were blue collar. By 1990 they were a distinct minority. This shift in the relative value of brain and brawn has worked to the advantage of women, and undoubtedly accounts for some of shrinkage in the gender wage gap. The left-panel of figure 1.22 shows women's participation in blue- and white-collar occupations as a fraction of all non-farm jobs (across both females and males). As can be seen, the majority of the increase in female employment has been in white collar jobs.

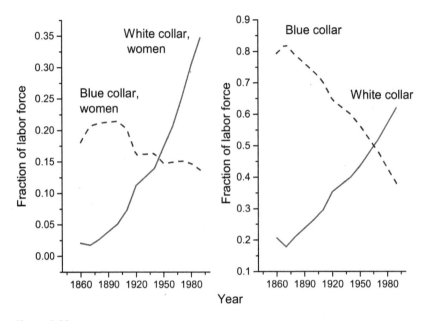

Figure 1.22
Blue- and white-collar jobs, 1860–1990. The right-hand-side panel shows the fraction
of blue- and white-collar jobs as a fraction of all non-farm jobs. The left-hand-side
panel plots women in blue- and white-collar jobs as a fraction of all non-farm jobs
(across both females and males).
Source: Calculated from *Historical Statistics of the United States: Millennial Edition* 2006,
tables Ba1033–1046 and Ba1061–1074, Cambridge University Press.

8.1   The Setting

Consider an economy inhabited by a representative married couple. The
married couple's preferences are given by

$$\theta \ln c + (1 - \theta) \ln(1 - h), \text{ with } 0 < \theta < 1,$$

where c is the household's consumption of goods and $1 - h$ are the hours
the woman devotes to home. Now, suppose that two types of labor input—
brain and brawn—are used in the production of goods. Both husband and
wife have one unit of time. The husband uses his one unit of time to supply,
inelastically, one unit of brain *and* one unit of brawn. The wife has only
one unit of brain. She can divide her unit of time between staying at home,
$1 - h$, or to supplying brains, h. This is not to suggest that women are inca-
pable of manual labor. In times past people of both genders did a lot of
physical labor. Rather it is an abstraction to simplify the analysis. All that

is really needed is the assumption that men (women) have a comparative advantage at doing manual (mental) labor.

Goods are produced in the economy using two production processes. The first process is a primitive technology and uses just brawn. Here physical labor (brawn), h_p, can be employed to produce output, o_p, in line with the production function

$$o_p = \pi h_p.$$

The second process is more advanced and uses only mental labor (brains), h_m, to produce output, o_m. Here,

$$o_m = \beta h_m.$$

Given the linear production structure, the wage rate for a unit of brawn is π and for a unit of brain is β. This transpires because in equilibrium the wage rate for brawn (brain) will equal its marginal product, π (β). The household's budget constraint reads

$$c = \pi + \beta + \beta h = \pi + \beta(1 + h),$$

where the right-hand side is the combined market earnings of the man, $\pi + \beta$, and the woman, βh.

8.2   The Household's Choice Problem

The household's decision problem for how much the woman should work in the market is

$$\max_{0 \le h} \{\theta \ln[\pi + \beta(1 + h)] + (1 - \theta) \ln(1 - h)\},$$

which has the first-order condition

$$\theta \underbrace{\frac{\beta}{\pi + \beta(1 + h)}}_{\text{MB}} \le (1 - \theta) \underbrace{\frac{1}{1 - h}}_{\text{MC}}.$$

There are two potential solutions to this first-order condition, an interior solution, $h \ge 0$, and a corner one, $h = 0$. This first-order condition holds with equality when an interior solution exists. In this situation, the solution for market hours, h, equates the marginal benefit and marginal cost from working an extra unit of time in the labor market. The interior solution for h is given by

$$h = \theta - (1 - \theta)\frac{(\pi + \beta)}{\beta}.$$

It is easy to deduce from the interior solution that $h \geq 0$, if $\beta/(\pi + \beta) \geq (1-\theta)/\theta$. At the corner solution, $h=0$, the marginal marginal benefit from working is less than its marginal cost. So, the woman will not work.

The full solution for h reads

$$h = \begin{cases} \theta - (1-\theta)\frac{(\pi+\beta)}{\beta} \text{ (work)}, & \text{if } \phi = \beta/(\pi+\beta) \geq (1-\theta)/\theta; \\ 0 \text{ (don't work)}, & \text{if } \phi = \beta/(\pi+\beta) < (1-\theta)/\theta. \end{cases}$$

Now, when a woman works in the market she will earn the wage rate $w_f = \beta$, while a man gets $w_m = \pi + \beta$. Thus, the gender wage gap is given by $\phi = \beta/(\pi + \beta) < 1$. The woman works when the gender wage gap, $\phi = \beta/(\pi + \beta)$, is greater than or equal to $(1-\theta)/\theta$, and doesn't do so otherwise. So she is more likely to work the less valuable brawn is relative to brain since $\beta/(\pi + \beta) = 1/(\pi/\beta + 1)$. Additionally, she is more likely to work the lower the utility weight on her time spent at home is relative to consumption, $(1-\theta)/\theta$. When she does work, her time spent in the market increases in the gender wage gap term, $\phi = \beta/(\pi + \beta)$.

8.3   The Rise in Married Female Labor-Force Participation and the Gender Wage Gap

How will industrialization affect the gender wage gap and married female labor-force participation? Go back to yesteryear when brain was not so important. This would imply a low value for β. Clearly, when β is small enough (think $\beta \simeq 0$) married women will not work since the gender wage gap, ϕ, will be close to zero. With economic development and the introduction of modern machinery, brawn is devalued; i.e., π drops. Similarly, tasks in the modern economy are more analytically based, which gives more currency to brain; that is, β increases. Thus, the gender wage gap, $\phi = \beta/(\pi + \beta) = 1/[\pi/\beta + 1]$, will move up with economic development. Married women will enter the labor force as ϕ rises. Over time they will increase their hours as well, since h is increasing in ϕ. Hence, the gender wage gap will shrink (or ϕ will rise) at the same time as married female-labor force participation increases. All of this occurs because technological progress in the economy favors jobs in which women have a comparative advantage.

9   The Division of Labor in the Household

Historically married women spent more time at home and less time working in the market than did married men. Even today this is true, as the
facts in section 1 of this chapter firmly establish. This fact motivated the
assumption that men always work full time in the market, which makes
the analysis of married female labor supply much easier. The assumption
that men work a fixed amount of time in the market is relaxed now. In
particular, the household is now free to pick the market hours for both the
man and woman. The analysis below is also germane for cohabiting couples
(and for the most part same-sex ones too). It will be shown that, in general,
it pays for at least one of the parties in a relationship to specialize in either
market work or household work.

9.1   A Framework for Analysis

Suppose that men and women each have one unit of time that can be freely
divided between market and nonmarket work. The market wage rate is
assumed to be w for an efficiency unit of labor. Let a man have the productivity μ_m in the market, so that a unit of his time there can supply μ_m
efficiency units of labor. A unit of his time at home supplies v_m efficiency
units of labor. The analogous datums for a woman are μ_f and v_f. (For same-
sex couples let the subscript m refer to one of the partners and f to the
other.) The household's utility function is given by

$$u = \theta \ln c + (1 - \theta) \ln n, \text{ with } 0 < \theta < 1,$$

where c and n are its consumption of market and nonmarket goods, respectively. Nonmarket goods, n, are produced according to the linear household
production function

$$n = v_m l_m + v_f l_f,$$

where l_m and l_f are the man's and woman's time spent in housework.
The household's income from the market is $w\mu_f(1 - l_f) + w\mu_m(1 - l_m)$.

9.2   The Household's Decision Problem

The household's maximization problem is

$$\max_{0 \le l_f, l_m \le 1} \{\theta \ln[w\mu_f(1 - l_f) + w\mu_m(1 - l_m)] + (1 - \theta) \ln(v_f l_f + v_m l_m)\}.$$

Assuming interior solutions, the two first-order conditions for l_f and l_m are

$$\theta \frac{w\mu_f}{w\mu_f(1-l_f)+w\mu_m(1-l_m)} = (1-\theta)\frac{v_f}{v_f l_f + v_m l_m}, \tag{9.1}$$

and

$$\theta \frac{w\mu_m}{\mu_f w(1-l_f)+w\mu_m(1-l_m)} = (1-\theta)\frac{v_m}{v_f l_f + v_m l_m}. \tag{9.2}$$

The left-hand sides of these first-order conditions represent the marginal cost of housework, while the right-hand sides portray the marginal benefits. Dividing the second first-order condition into the first gives

$$\frac{\mu_f}{\mu_m} = \frac{v_f}{v_m}, \text{ or } \frac{\mu_m}{v_m} = \frac{\mu_f}{v_f}.$$

This implies that two interior solutions can obtain only when $\mu_m/v_m = \mu_f/v_f$, or when the relative productivities of market work and housework are the same for the man and woman. Otherwise, one or both of them should specialize in a single type of work.

9.3   Who Should Work Where?

Now, suppose that $\mu_m/v_m > \mu_f/v_f$. That is, the man has a comparative advantage in market work. Therefore, one would expect him to spend a higher fraction of his time in market work than his wife. Three cases can happen. First, $0 < l_f < 1$ and $l_m = 0$: here the man has completely specialized in market work, and the women works in the home and in the market. Second, $l_f = 1$ and $l_m = 0$: as before the man has completely specialized in market work, but now the woman is completely specialized in home work. Third, $l_f = 1$ and $0 < l_m < 1$: in this situation the woman is completely specialized in home work, and the man does some housework in addition to working in the market. In all cases at least one person is completely specialized.

Start with the first case. If equation (9.1) holds with equality, then there will be an interior solution for the woman's time at home, l_f. In this situation, the left-hand side of (9.2) must be bigger than the right-hand side because $\mu_m/v_m > \mu_f/v_f$. That is, the marginal cost for a man to do housework exceeds its marginal benefit. Thus, the man must be at corner solution where he devotes all of his time to the market and none to housework; i.e.,

$l_m = 0$. (The mathematics associated with a corner solution are discussed in the Mathematical Appendix.) Using this fact in (9.1) yields

$$\theta \frac{\mu_f}{\mu_f(1-l_f)+\mu_m} = (1-\theta)\frac{1}{l_f},$$

which gives the woman's interior solution for housework

$$l_f = (1-\theta)(1+\mu_m/\mu_f).$$

As can be seen, she will spend less time at home (and more in the workplace) the higher her productivity in the market, μ_f, is relative to her partner's, μ_m. Since l_f is bounded above by one, this interior solution holds only when $(1-\theta)(1+\mu_m/\mu_f) \le 1$, or equivalently when $\theta/(1-\theta) \ge \mu_m/\mu_f$. Otherwise, $l_f = 1$. Taking stock of the situation, the first case transpires when

$$\frac{\theta}{1-\theta} \ge \frac{\mu_m}{\mu_f} > \frac{v_m}{v_f}.$$

In this case market goods have a high relative value in the sense that $\theta/(1-\theta)$ is large relative to μ_m/μ_f. So, the woman should spend some time in the market.

To address the second case, suppose that the left-hand side of (9.1) is less than the right-hand side one, evaluated at $l_f = 1$ and $l_m = 0$. Then, the marginal cost of the woman's doing housework is less than its marginal benefit. For this to happen, it must transpire from (9.1) that $\theta/(1-\theta) < \mu_m/\mu_f$. Additionally, for $l_m = 0$ the left-hand side of (9.2) should be greater than the right-hand side when $l_f = 1$ and $l_m = 0$. This occurs when $\theta/(1-\theta) > v_m/v_f$. So, the condition for the second case is

$$\frac{\mu_m}{\mu_f} > \frac{\theta}{1-\theta} > \frac{v_m}{v_f}.$$

Compared with the first case, market goods are less valuable in the sense that $\theta/(1-\theta)$ is small relative to μ_m/μ_f. Now, the women should specialize in home work.

Last, for the third case, note that it is possible that (9.2) holds with equality when the left-hand side of (9.1) is less than the right-hand side. Here it will transpire that $l_f = 1$ and $0 < l_m < 1$. By setting $l_f = 1$ in (9.2), the interior solution for the man's time spent at home is given by

$$l_m = (1-\theta) - \theta v_f/v_m.$$

For this solution to be valid, l_m must be nonnegative, which implies that the restriction $\theta/(1-\theta) \leq v_m/v_f$ must hold. The third case occurs when

$$\frac{\mu_m}{\mu_f} > \frac{v_m}{v_f} \geq \frac{\theta}{1-\theta}.$$

Compared with the second case, market goods have a low relative value in the sense that $v_m/v_f \geq \theta/(1-\theta)$. Here the man should spend some time on housework.

Last, it is *not* possible to have both parties specialize in the same type of work. That is, the situations where either $l_f = l_m = 0$ or $l_f = l_m = 1$ cannot occur. If one of these cases did, then either the consumption of market or home goods would be zero. This is ruled out with a logarithmic utility function because household utility would then be $-\infty$.

The gender wage gap may have implied that, historically speaking, $\mu_m > \mu_f$. This could have happened because the brawn component of market jobs was high compared with today. Additionally, the biology of bearing and raising young children (say, due to necessities such as breastfeeding, which is discussed in chapter 2) could have resulted in $v_f > v_m$. These factors would then have led to men specializing in market work, as the above example shows. When μ_m/μ_f is very high, women would spend all of their time at home. As this falls, due to economic development, women would start to work in the market. If $\mu_m/v_m = \mu_f/v_f$, then it doesn't matter who works where. The couple could do both home and market work. In modern times, for some households $\mu_m/v_m < \mu_f/v_f$; i.e., the woman has the highest relative productivity for market work. And one does see situations where the woman works full time in the market and the man does the housework.

10 Literature Review

The economic analysis of female labor-force participation began with the pioneering works of Mincer (1962) and Cain (1966). An elementary introduction to the economics of women in the workplace is contained in Blau, Ferber, and Winkler (2014). The massive rise in female labor-force participation over the course of the twentieth century has attracted a lot of notice from economists, who have examined the extent to which the rise in real wages and the narrowing of the gender wage gap can account for the rise in labor-force participation. The discussion of the gender wage gap in section 7 of this chapter draws upon Blau and Kahn (2008, 2017), who survey the research on this topic. Blau, Ferber, and Winkler (2014) discuss the occupational differences between men and women, upon which this section also relies. A well known book on the gender wage gap is by Goldin (1990).

Galor and Weil (1996) provide an interesting general equilibrium model in which the increase in women's wages and labor-force participation is a byproduct of the process of development. Here capital accumulation in the market sector raises women's wages relative to men's wages. The underlying mechanism is that capital in the market sector is more complementary to women's labor than it is to men's labor, since it displaces the need for physical strength. The analysis in section 8 is inspired by Galor and Weil (1996). The increase in women's wages, associated with the transition from brawn to brain, also raises the cost of children in their model. This causes a drop in fertility. (See chapter 2 for a discussion about how raising wages increase the cost of having children.) Cortes, Jaimovich, and Siu (2017) document the recent rise of women and the decline of men in high-wage cognitive occupations. Using data from the U.S. Department of Labor's *Dictionary of Occupational Titles* on the skill and temperament traits required to perform tasks linked with various occupations, they find that high-wage cognitive jobs increasingly demand social skills. Research in neuroscience and psychology indicates that women have a comparative advantage in tasks requiring interpersonal and social skills. Jones, Manuelli, and McGrattan (2015) argue that increases in the gender wage gap, $\phi = w_f/w_m$, can account for increases in average hours worked by married women for the time period between 1950 and 1990.

Fernandez, Fogli, and Olivetti (2004) present evidence suggesting that a man is more likely to have a working wife if his own mother worked than if she didn't. In particular, men who had mothers who worked during World War II had a higher likelihood of marrying working women than those who didn't. Fernandez, Fogli, and Olivetti (2004) develop a model in which attitudes toward working women become more receptive over time. A version of the model is presented in chapter 4.

The economic importance of household production was probably first recognized in a classic book by Reid (1934). She carefully reported and analyzed the uses of time and capital by households of the era. The data was fragmentary then, and still is. Reid (1934) knew in theory that labor-saving household capital could reduce the amount of time spent on housework, but the evidence that was just emerging at the time suggested that this effect was modest (see table 13, 91). From the Aguiar and Hurst (2007) and Gershuny and Harms (2016) studies, it is now known beyond doubt that the time spent on household work declined dramatically over the course of the entire twentieth century. Furthermore, adults spend less time on cooking and cleaning in richer countries than in poorer ones, as Bridgman, Duernecker, and Herrendorf (2017) document, so time spent on housework falls with economic development.

In a famous paper Becker (1965) develops the modern approach to household production: the treatment of the household as a small factory or plant using inputs, such as labor, capital, and raw materials, to produce home goods. He discusses, ever so briefly, the division of labor within the household. Recall that Becker (1965) dismissed the concept of leisure, implying that it is arbitrary in nature. Still, one could add it into the analysis by subdividing nonmarket time into home work and leisure time. Gronau (1986) operationalizes the definition of leisure. He defines work at home to be those activities one could hire someone else to do and suggests that it is impossible to enjoy leisure vicariously. The Aguiar and Hurst (2007) study illustrates how nonmarket time can be broken down into home work and leisure time, although there is some arbitrariness about what constitutes leisure as reflected by their four measures. Ermisch (2003) has a chapter on household production, in which he discusses the division of labor within the household.

According to Black, Sanders, and Taylor (2007) about 32 percent of traditional families in 2000 had a stay-at-home partner. The partner who stays at home is less likely to have a college degree than the one who works. The stay-at-home partner is much more likely to be the woman (71 percent in 2000). Most families with stay-at-home women have children (70 percent). In typical traditional families where both partners work, the man earns more than the woman (67 percent), and he puts in more hours in the market than she does (45.9 vs 35.5 hours per week). It is interesting to look at the statistics for same-sex couples, since there are no biological differences between the partners. A smaller fraction of gay and lesbian families have a stay-at-home partner (roughly 19 percent do). Here, too, a smaller percentage of the partners who stay at home have a college degree than the ones that work. Moreover, same-sex couples with a stay-at-home partner are much less likely to have children (only 32 percent do for lesbian families and 16 percent for gay ones). In gay and lesbian families where both partners work the one who earns the most works somewhat more (about 4 hours a week more), but the difference is not nearly as pronounced as in traditional families.

Benhabib, Rogerson, and Wright (1991) and Greenwood and Hercowitz (1991) introduce household production theory into dynamic general equilibrium models in order to study the movement of labor over the business cycle. The idea is that in favorable economic times, households may temporarily move labor out of the home sector to take advantage of good market opportunities, thereby increasing the elasticity of labor supply. Parente, Rogerson, and Wright (2000) use a similar framework to investigate whether household production can explain cross-country income differentials. Rios-Rull (1993) inserts household production into an overlapping generations model to examine its impact on the time allocations of skilled versus unskilled labor. In his framework, skilled labor (relative to unskilled labor) tends to substitute market goods or services for labor in household production.

Greenwood, Seshadri, and Yorukoglu (2005) illustrate how labor-saving appliances and intermediate goods can encourage married female laborforce participation. They embed a Becker-Reid household production model into a dynamic general equilibrium model. Parts of the historical discussion have been drawn from there, as well as from Greenwood, Seshadri, and Vandenbroucke (2005). The underlying data sources for figures 1.13, 1.14, and 1.15 are detailed in Greenwood, Seshadri, and Yorukoglu (2005). Empirically, female labor supply is negatively associated with the price of household appliances, as shown by Cavalcanti and Tavares (2008) in a panel of OECD countries. Heisig (2011) looks at a wider set of countries and, using a direct measure of appliance diffusion, finds that household technologies are an important force behind increasing female labor-force participation. Likewise, using U.S. Census micro data, Coen-Pirani, Leon, and Lugauer (2010) document that a significant portion of the rise in married female labor-force participation during the 1960s can be attributed to the diffusion of household appliances. Similar evidence is provided by Dinkelman (2011), who studies the effect of rural electrification in South Africa. In addition, Albanesi and Olivetti (2016) argue that advances in maternal medicine also reduced the need for a married woman to stay

at home. Some of Albanesi and Olivetti's (2014) facts are presented in
chapter 2.

Last, calibration is discussed in Prescott and Candler (2008).

11 Problems

(1) *The decline in men's market hours, theory and calibration.* Take the unisex
single model of labor supply. Suppose that the household's preferences are
given by

$\theta \ln(c - \psi) + (1 - \theta) \ln l$, with $0 < \theta < 1$ and $\psi \geq 0$.

Here $\psi \geq 0$ is a constant. It represents a subsistence level of consumption for
the household. That is, consumption, c, must be a least as large as ψ
for the household to enjoy any utility. The household has one unit of time.
It divides this up between working in the market, $1 - l$, at wage w, and
leisure, l.

(a) Set up and solve the household's maximization problem.

(b) How does a rise in the wage rate, w, affect time spent in the market
and leisure? Translate your mathematics into a graph. Intuitively, how does
ψ work?

(c) Apply the model to a household made up of a lone male. Suppose that
in 1900 a man worked 70 hours a week. Set the wage rate in 1900 to 1.
(Explain why this can be done.) Assume that a male's labor had dropped to
42 hours per week by 2000 and that the wage rate was 9 times larger. Is the
unitary model capable of explaining these facts? Show why or why not.

(2) *Advances in maternal medicine.* Most married women have children.
Given considerable advancements in maternal medicine over the last 100
years, it has become much easier for women to return to work after giving
birth. Before the twentieth century many women died from hemorrhage,
sepsis, toxemias, and traumatic incidences associated with childbirth. (This
is discussed in chapter 2). Often women endured a long period of mor-
bidity after childbirth. The introduction of drugs, such as penicillin and
sulfonamides, helped to alleviate all of this. Furthermore, baby formula
eased the burden of breastfeeding. By comparison, around 1900 it was hard
for a married woman to go to work.

Consider the following model of married female labor supply along the
extensive margin. A woman's labor, h_f, is constrained to lie in the two-point

set, $\{0, \bar{h}_f\}$. A women earns the wage, w_f, and a man, w_m. A man's labor supply is fixed at $\bar{h}_m$. The household has preferences of the following form,

$$u = \theta \ln c + (1 - \theta)\xi \ln(1 - h_f) + (1 - \theta) \ln(1 - \bar{h}_m),$$

with $0 < \theta < 1$. The variable ξ denotes the value that the couple places on the woman's time spent at home. The couple enjoys all utility flows equally. Assume that all married women give birth to a pair of twins. Let π represent the probability that everything will go smoothly. In this case, $\xi = \lambda$. Let $1 - \pi$ be the odds that a woman will experience some difficulty with the delivery of her twins. In this case, $\xi = \kappa\lambda$ where $\kappa > 1$. Let λ be distributed in line with the following Pareto distribution:

$$\Pr[\lambda \leq x] = 1 - \left(\frac{1}{x}\right)^\gamma, \text{ with } x, \gamma \geq 1.$$

Note that a woman decides to go to work *after* she has given birth.

(a) Set up and solve the household's maximization problem.

(b) Characterize how many women will work in the economy. What happens as κ falls? Explain the economics behind your result.

(c) What is the lowest value for κ that would imply that all sick women don't work?

(d) In 1900 just 5 percent of married women worked. This rose to 50 percent by 1980. Find the values of γ and π that are consistent with these facts. When doing this, assume that a man spends 40 percent of his time working while a working woman spends 25 percent. Suppose that a woman earns 75 percent of what a man does. Assume that in 1900 no sick women worked, and that κ is 1 in 1980. Set $\theta = 0.5$.

2 The Baby Boom and Baby Bust

This chapter aims to explain two key facts: first, the long-run decline in fertility that characterized most Western countries in the nineteenth and twentieth centuries; second, the boom in fertility around World War II that temporarily interrupted the long-run decline. The hypothesis advanced here is that these two phenomena are the result of technological progress in the market and at home. The long-run decline in fertility is explained by the secular increase in wages, which raised the opportunity cost of having children. The spread of labor-saving technologies in the home and the advance of maternal medicine are put forth as drivers of the baby boom. Fifty years ago it would have seemed strange to think that standard economic theory can be used to analyze fertility.

The chapter starts off with some commonly used measures of fertility. It then moves on to present some facts (using these measures) about the secular decline in fertility and the baby boom. The conventional view that the baby boom was a catch-up effect in fertility due to the return of soldiers absent during World War II is debunked. After this, a model of the baby boom and baby bust is developed. This baseline model uses rising wages to explain the baby bust and the introduction of labor-saving household technologies in order to rationalize the baby boom. The framework is then extended to incorporate advances in maternal medicine, which serves as a complementary hypothesis for the baby boom.

Three interesting side topics will also be pursued. First, prior to World War II, major conflicts were connected with large drops in fertility. For World War I France, it has been estimated that the loss in unborn babies was about as large as the number of war deaths. The commonplace mechanical explanation is that this was simply due to an absence of men. Perhaps, but given the large number of male casualties and deaths, there would still be an absence of men after the war when fertility displayed a strong rebound. An alternative economic explanation is advanced here: the expected cost of having children during a major conflict is high. The second topic concerns the fact that women need to take time off to bear and raise children. This may influence the type of jobs that they take, as well as their educational preparation. The pay that they earn relative to men will also be affected and have implications for the gender wage gap. Over time, as fertility drops, the tradeoff between kids and jobs may not be as important. The third topic focuses on the Malthusian model of population determination. Malthus believed that the size of the population was limited by the productive capability of land, a technological factor. Many believe that this theory is a good description of the preindustrial era. Some historical evidence is presented for the period prior to the Industrial Revolution in England. The last section discusses some of the literature relevant for the chapter.

1 Definitions of Fertility

There are many concepts that can be used to measure fertility within a country. Five widely used measures are:

(1) *Crude Birthrate*. This is simply defined as

$$\text{CRUDE BIRTHRATE} = 1{,}000 \times \frac{\text{total number of births}}{\text{total population}}.$$

It is a rough measure of the number of births that 1,000 people have in a year. Figure 2.1 plots crude fertility rates for four regions of the world. The striking fact is that in modern times fertility has declined rapidly all over the world. The chart for Asia has a ∩-shaped pattern that is characteristic of economic development. At the early stages of economic development fertility tends to be low. A household cannot afford a large number of children. Fertility then rises as a country gets richer and households can afford larger families. Eventually, at higher stages of development, fertility starts to fall with income, perhaps because women enter the labor force and/or because parents want to focus on having fewer kids with better educations. This ∩-shaped pattern is called the *demographic transition*. The timing of

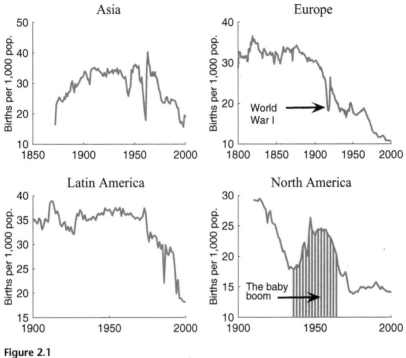

Figure 2.1
Demographic transitions.
Source: Greenwood, Guner, and Vandenbroucke 2017.

demographic transitions varies across regions of world because some countries developed earlier and faster than others. So, for instance, England had a demographic transition relatively early on, with the peak of the ∩ being around 1800. In Asia the shift happened much later because economic development was farther behind. Also note the sharp drop in European fertility (about 50 percent below trend) during World War I. Wars and revolutions were often met with drops in fertility, as discussed later on. When looking at fertility rates, it may be more relevant to focus on the fecund populace—i.e., those women who are capable of giving birth. This leads to the next measure.

(2) *General Fertility Rate*. This is given by

$$\text{GENERAL FERTILITY RATE} = 1{,}000 \times \frac{\text{total number of births}}{\text{women aged 15–49}}.$$

The crude and general fertility rates are sensitive to the age structure of a population. Younger woman have a higher fertility rate than older ones. The next measure adjusts better for this.

(3) *Age-Specific Fertility Rates*. One can measure fertility for women in certain age ranges using the age-specific fertility rate.

AGE-SPECIFIC FERTILITY RATES (for women of age group i) $= f_i$

$$= 1,000 \times \frac{\text{total number of births to women in age interval } i}{\text{women in age interval } i}.$$

So, for example, one could look at the fertility rate for women aged 20–24 or 35–39. Age-specific fertility rates can be used to compute a measure of aggregate fertility in a way that controls for the age structure of the population, as the next measure indicates.

(4) *Total Fertility Rate*. The total fertility rate attempts to measure the number of children that a representative woman will have over her lifetime starting in a given year. It is defined by

$$\text{TOTAL FERTILITY RATE} = \sum_i \frac{I_i f_i}{1,000},$$

where I_i is the length of the age interval in years for each age group category i and f_i is the age-specific fertility rate for women in age group i. The age groups could be women aged 15–19, 20–24, 25–29, 30–34, 35–39, and 40–44, for example. The age-specific fertility rates are taken from a cross-section of women of various ages at a single point in time. So, the total fertility rate will not truly give a woman's lifetime fertility unless the cross-sections for fertility are stable over time. Since the age-specific fertility rate, f_i, is annual, in order to get the number of births over a woman's lifetime, one must multiply the age-specific fertility rate by the length of the age interval for the age category, I_i. Additionally, age-specific fertility rates are measured per 1,000 people so the total fertility rate needs to be divided by this number. The end result is the projected number of births over a woman's life.

(5) *Completed Fertility Rate*. This calculates the number of children who were actually born per woman in a cohort of women up to the end of their childbearing years. Clearly, this can be done only for older women who have passed their childbearing years. The completed fertility rate is excellent for historical research.

2   Fertility, 1800–1990

The fertility of American women over the last two hundred years has two salient features, portrayed in figure 2.2. First, it has declined drastically. The completed fertility rate for the average white woman in 1800 was 7 children. By 1990 this had dropped to just 2. It would be hard to attribute this decline to the introduction of modern contraceptives, such as the pill. The general fertility rate dropped from 278 births per 1,000 people in 1800 to just 67.2 in 1989. However, it had already fallen to 130, by 1900—well before modern contraceptives. Moreover, the rate was at 118 in 1960, which is before the introduction of the pill. (A time line on technological progress in contraception is presented in chapter 4.) The decline in fertility was unabated during the 140 year period between 1800 and 1940.

Second, fertility showed a surprising recovery between the mid-1940s and mid-1960s. The upturn was large, a "baby boom."[1] Just how large depends upon the concept of fertility used. For example, the general fertility rate increased by 41 percent between 1934 and 1959. Alternatively, the completed fertility rate was 28 percent higher for a woman born in 1932 (whose average child arrived in 1959) than for one born in 1907 (whose average child was born in 1934).[2] The difference between these two numbers suggests that the rise in fertility was compressed in time for two reasons. First, as will be seen, older women had more children. Second, so did younger women. But, the high rates of fertility that younger women had early in their lives were not matched by higher rates of fertility later on. This leads to the last point. After the mid-1960s fertility reverted back to trend—which is to say that the "baby bust" resumed.

2.1   The Cross-Sectional Relationship between Fertility and Income
One hypothesis suggested by figure 2.2 is that the secular decline in fertility is due to a rise in income. If this is true, then at any point in time richer families should have fewer children than poorer families. So, what does the

1. The baby boom in the United States is conventionally dated as occurring between 1946 and 1964. As will be seen, these dates are suspect.
2. Note that the average childbearing age was 27, roughly the horizontal distance between the two curves. The completed fertility rate is plotted for the years when the mothers were born, while the general fertility rate is charted for the years when the babies were born.

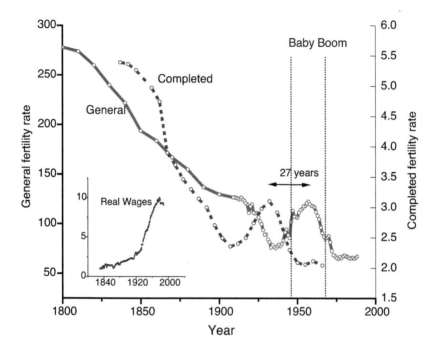

Figure 2.2
Fertility in the United States, 1800–1990.
Source: Greenwood, Seshadri, and Vandenbroucke 2005.

cross-sectional relationship between fertility and income look like? Jones and Tertilt (2008) have investigated this question. Some of their evidence is presented in figure 2.3, which plots completed fertility for women born in a particular year (their cohort group) against family income, measured in 2000 dollars. Twenty such cross-sections are examined by Jones and Tertilt (2008). Although family income is not available for the early years in the U.S. census, the occupation of the husband is. Jones and Tertilt (2008) therefore construct an index of family income using data on occupations. As can be seen from figure 2.3, for any cohort the cross-sectional relationship between fertility and income is downward-sloping. Thus, richer families do have fewer children. Yet, as one moves through time, the cross-sectional relationship moves about somewhat. For example, the profile for women born in 1938, who gave birth in the baby boom years, lies above the profiles for women born in 1878 and 1958, at least for the income observations

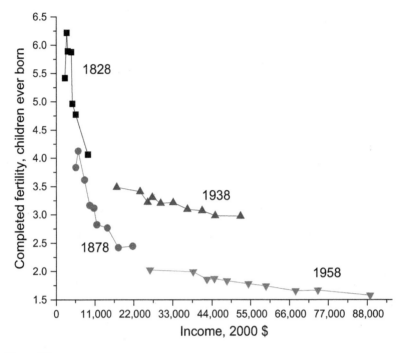

Figure 2.3
Fertility cross sections at different points in time for the United States. The year denotes the birth date for each cohort of women.
Source: Jones and Tertilt 2008, figure 4.

that overlap. Is there a stable relationship between fertility and income over time?

According to Jones and Tertilt (2008), about 84 percent of the decline in fertility between 1828 and 1958 can be predicted by using the cross-sectional relationship between income and fertility for the 1828 cohort. Imagine fitting a relationship, or regression, of the form

$$\ln(\text{FERTILITY}) = \text{CONSTANT} + \beta \ln(\text{INCOME}),$$

to the data for the 1828 cross-section. You will find that $\beta = -0.33$, which represents the income elasticity for fertility. Using the estimated 1828 cross-sectional relationship, you can make an estimate of what fertility should have been for the 1958 cohort of women. Between 1828 and 1958, income increased by 13 times from \$4,154 to \$54,517. The above relationship implies that

$$\text{FERTILITY}_{1958} = \left(\frac{\text{INCOME}_{1958}}{\text{INCOME}_{1828}}\right)^{-0.33} \times \text{FERTILITY}_{1828}$$

$$2.4 = \left(\frac{\$54,517}{\$4,154}\right)^{-0.33} \times 5.6.$$

The actual value for fertility in 1958 was in fact 1.8. Hence, the regression predicts a drop of 3.2 (or 5.6 − 2.4) kids per woman versus the realized fall of 3.8 (or 5.6 − 1.8). This is roughly 84 percent of the observed decline. So, in this sense, the relationship between fertility and income does appear to be constant over time.

3   The Mystery of the Baby Boom

Conventional wisdom links the baby boom with the end of the Great Depression and World War II. The popular view is that these traumatic events led to a drop in fertility. Part of the decline in fertility was due to economic hardship or a gloomy outlook on the future, which made it difficult to start a family. Part of it was due to the absence of so many young men, who had gone off to fight the war. After World War II fertility bounced up, as the men returned, the economy boomed, and a general feeling of optimism prevailed. Fertility rose to above-normal levels to make up for lost fertility during the Depression and war years. This explains the baby boom, according to conventional wisdom.

Conventional wisdom is often wrong. Specifically, the pattern and timing of fertility do not support the belief that the baby boom was the outcome of the Great Depression and World War II. In particular:

(1) Figure 2.2 shows *completed* fertility for the women who gave birth during the baby boom. A pure catch-up effect should have had no influence on lifetime fertility, since one less child today would just be made up for by having one more child tomorrow. Yet, lifetime fertility rose.

(2) Take the peak of the baby boom for the United States, the year 1960. As will be discussed below, the cohort of women contributing the most to the baby boom then (those in the 20–24 age group) were simply too young for either the Great Depression or World War II to have had much of an impact on them. They were not alive during the Great Depression, and were less than 9 years old at the end of World War II.

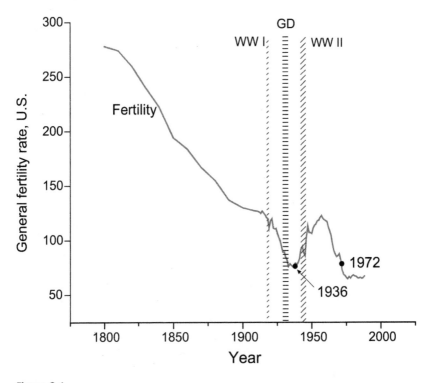

Figure 2.4
U.S. fertility, 1800–1990.
Source: Greenwood, Seshadri, and Vandenbroucke 2005.

(3) The data show that the baby boom actually started in the 1930s for the United States and many countries in the Organisation for Economic Co-operation and Development (OECD). This will be detailed below. Furthermore, for many countries fertility *grew* throughout World War II, and the neutral countries, Ireland, Sweden, and Switzerland, all had baby booms. In addition, it's hard to detect a precipitous drop in the United States and many other OECD countries' fertilities due the Great Depression.

So, how does conventional wisdom about the baby boom match up with the pattern of fertility displayed in the U.S. data, as shown in Figure 2.4? Observe that the general fertility rate fell continuously from 1800 to about 1936. It then began to rise. One interpretation of the graph is that the baby boom started in the 1930s. The upward trend suffered a slight drop from 1943 to 1945 during World War II. Note that fertility fell during World War I and then rebounded. Additionally, note that there is no unusual decline in fertility associated with the Great Depression (GD). A nondemographer eyeballing this graph might date the baby boom as occurring from 1936 to 1972.

Could the baby boom be some sort of catch-up effect associated with World War II? First, if the baby boom were merely the result of couples postponing family formation during the war years, then there should be no increase in completed fertility (or lifetime births) for a woman. Yet, completed fertility did increase for fecund women during the baby boom years, as figure 2.2 shows. Second, many of the women giving birth during the baby boom were simply too young for such a catch-up effect to be operational. Figure 2.5 plots fertility rates for various age groups of white women. These age-specific fertility rates are weighted by the relative size of each group. Therefore, the diagram provides a measure of the contribution of each age group to the baby boom. Take the 20–24 age group: They contribute the most to the baby boom. This series peaks in 1960. But at the peak, the members of this group were somewhere between 1 and 5 years old in 1941 and 5 and 9 years old in 1945. Hence, a catch-up effect is impossible for them. It's also implausible that the Great Depression affected their fertility decision. Fertility for the 25–29 age group rises until about 1952, then levels off until 1957, and declines thereafter. Those giving birth in 1952 would have been in the 14–18 age range in 1941 and in the 18–22 range by 1945, while those having kids in 1957 would have been somewhere between 9 to 13 years old in 1941 and 13 to 17 in 1945.

Some evidence of delayed fertility can, however, be gleaned from the diagram, but it looks small. For example, note that the fertility rate for the 20–24 age group starts to fall in 1942, as marked by point A. In 1947 these women would have been in the 25–29 age group. Note the small peak in 1947 for the latter age group, marked by point A'. Similarly, the fertility rate for the 15–19 age group begins to fall in 1943—point B. The majority of this group would have been in the 20–24 age group around 1948. Note the spike in 1947 for this age group, marked by point B'.

U.K. fertility dropped more or less unabated from 1876 to 1940, with one exception (see figure 2.6). There was a sharp decline and rebound associated with World War I. The United Kingdom suffered a prolonged depression during the interwar years. Again, it would be hard to argue that there was an unusual decline in fertility during these years. Interestingly, fertility rose

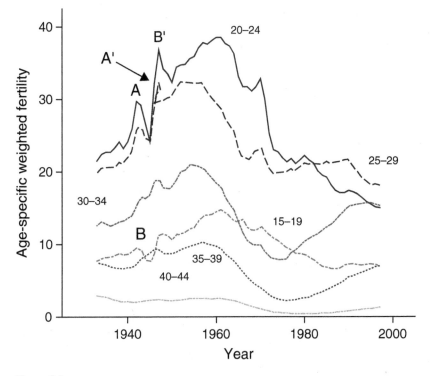

Figure 2.5
U.S. fertility by age group, 1933–1997.
Source: Greenwood, Seshadri, and Vandenbroucke 2005.

throughout World War II. The nondemographer might date the baby boom as occurring between 1941 and 1972. France shows a similar pattern with fertility rising throughout World War II (see figure 2.7). One might date the French baby boom as happening between 1942 and 1974. Last, even neutral countries, such as Switzerland, had a baby boom (see figure 2.8). Observe that Swiss fertility rose throughout World War II. Reasonable dates for the Swiss baby boom are 1937 to 1971.

4   A Basic Model of Fertility

It is interesting that standard economic theory can be used to analyze the determination of fertility, an issue that some might view as outside of the realm of economics. But a simple unisex model of fertility can be developed.

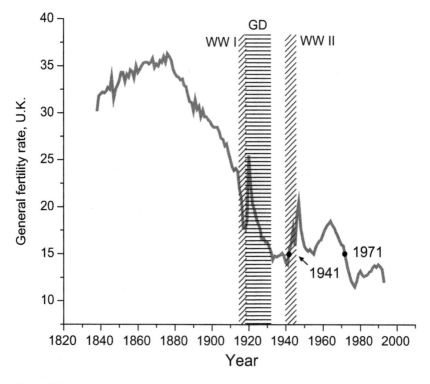

Figure 2.6
U.K. fertility, 1838–1993.
Source: Greenwood, Seshadri, and Vandenbroucke 2005.

Clearly, it is an abstraction: in the real world each child has two biological parents. In the unisex model reproduction is asexual: a lone parent can decide on the number of children to have, without having to consider a partner's opinion. This is a huge simplification since modeling children in two-parent family, where the partners have different views about the size of the family, may involve some notion of bargaining.

An adult has one unit of time, which can be split between working in the market or raising kids. A unit of time spent working earns the wage rate w. Suppose that a parent has tastes for consumption, c, and the number of kids, n, of the following form:

$$\theta \ln(c + \mathfrak{c}) + (1 - \theta) \ln n, \text{ with } 0 < \theta < 1, \tag{4.1}$$

where $\mathfrak{c} > 0$ is a constant. As will be seen, this constant plays an important role in the analysis. Think about it as representing some minimal

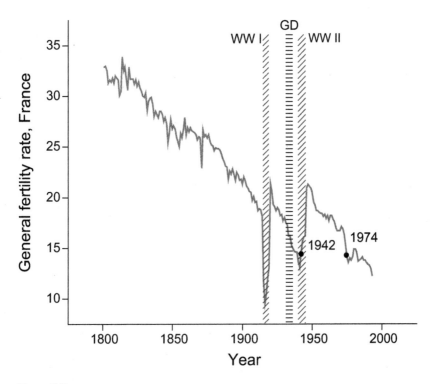

Figure 2.7
French fertility, 1801–1993.
Source: Greenwood, Seshadri, and Vandenbroucke 2005.

level of consumption that may arise from home production. Children provide utility, just like consumption (or leisure in a more general setting).

Raising children is costly, particularly in terms of parental time. Write a production function for children in the form

$n = xl^{1-\gamma}$, with $0 < \gamma < 1$,

where n is the output of children and l is the time spent rearing children. Envision x as representing the level of productivity in the home sector. The higher x is, the more children, n, can be raised for a given amount of time, l. As will be discussed, increases in x may reflect the introduction of labor-saving household technologies or advances in obstetric and pediatric medicine that reduced the time cost of having children. Since $1 - \gamma < 1$, this production function exhibits decreasing returns in l; that is, as l increases,

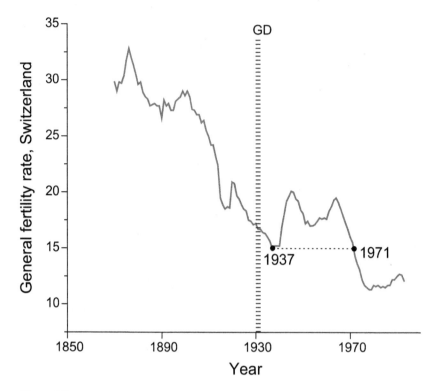

Figure 2.8
Swiss fertility, 1870–1993.
Source: Greenwood, Seshadri, and Vandenbroucke 2005.

each successive unit of time results in fewer extra kids. In other words, the production function is strictly concave in *l*. This production function can be inverted to give a cost function, expressed in units of time, for raising children, In particular, one gets

$$l = \left(\frac{n}{x}\right)^{1/(1-\gamma)}.$$

The time cost of raising children, *l*, is increasing in the number of children, *n*, and is decreasing in the productivity of the household sector, *x*. Last, because $1/(1-\gamma) > 1$, this cost function displays increasing marginal cost in *l*; i.e., as *n* rises, each extra kid costs increasingly more in terms of time. Hence, the cost function is strictly convex in *n*.

The parent's budget constraint can be written as

$$c = w(1-l).$$

By using the above cost function for l, the budget constraint can be expressed in terms of the number of kids, n:

$$c = w - w \left(\frac{n}{x}\right)^{1/(1-\gamma)}. \tag{4.2}$$

From the above equation, it is clear that the cost of raising children will be increasing in the wage rate, w. By having children, a parent takes time away from the market, $1 - l$, and this reduces consumption, c. It is also clear that the cost of rearing children is decreasing in household productivity, x.

The parent's maximization problem is now easily seen to be represented by

$$\max_{n} \left\{ \theta \ln \left[w - w \left(\frac{n}{x}\right)^{1/(1-\gamma)} + c \right] + (1 - \theta) \ln n \right\},$$

where consumption, c, has been solved out for using the budget constraint (4.2). The first-order condition associated with this maximization problem is

$$\underbrace{\frac{\theta}{w - w(\frac{n}{x})^{1/(1-\gamma)} + c}}_{\text{MU}_c} \times w \times \underbrace{\frac{1}{(1-\gamma)} \frac{n^{1/(1-\gamma)-1}}{x^{1/(1-\gamma)}}}_{\text{MC}_n} = \underbrace{\frac{1-\theta}{n}}_{\text{MU}_n}. \tag{4.3}$$

The right-hand side gives the marginal benefit from an extra child, which is just the marginal utility of a child, $\text{MU}_n = (1 - \theta)/n$. The left-hand side represents the marginal cost in terms of the foregone utility that arises from the drop in consumption associated with an extra child. This can be decomposed and explained as follows. The marginal time cost of a child is $\text{MC}_n = n^{1/(1-\gamma)-1}/[(1 - \gamma)x^{1/(1-\gamma)}]$. The parent could have worked instead, so this represents a loss in the amount $w \times \text{MC}_n = w \times n^{1/(1-\gamma)-1}/[(1 - \gamma)x^{1/(1-\gamma)}]$, in terms of foregone consumption. Since the marginal utility of consumption is $\text{MU}_c = \theta/(c + c) = \theta/[w - w(n/x)^{1/(1-\gamma)} + c]$, the left-hand side gives the utility cost of an extra child. So, the marginal utility cost of a child is $\text{MU}_c \times w \times \text{MC}_n$.

Fertility, n, will be affected by the wage rate, w, and the productivity of the home sector, x. To see how, solve the above equation for n. Doing this results in

$$n = \left[\frac{A}{1+A}\right]^{1-\gamma} x \left[1 + \frac{c}{w}\right]^{1-\gamma},$$

where $A \equiv (1 - \gamma)(1 - \theta)/\theta$. Two results can be seen from this equation:

(1) *Fertility, n, is decreasing in the wage rate, w, so long as $c > 0$.* This mechanism can generate a long-run decline in fertility. Recall that the marginal utility cost of having children is the product of three components, $MU_c \times w \times MC_n$. An increase in wages leads to a rise in opportunity cost of having children when measured in terms of consumption. That is, $w \times MC_n = wn^{1/(1-\gamma)-1}/[(1-\gamma)x^{1/(1-\gamma)}]$ rises. This can be thought of as representing a substitution effect. A rise in wages will also cause consumption to rise, which, in turn, leads to a drop in the marginal utility of consumption, $MU_c = \theta/[w - w(n/x)^{1/(1-\gamma)} + c]$. This is analogous to an income effect. Which effect will dominate? The above equation shows that the former effect does when $c > 0$. They cancel out when $c = 0$. When $c > 0$ (as opposed to $c = 0$) the marginal utility of consumption declines at a slower rate when consumption, $c = w - w(\frac{n}{x})^{1/(1-\gamma)}$, rises with w. To see this, think about increasing wages by a factor of λ. Whereas $w - w(n/x)^{1/(1-\gamma)}$ will rise by a factor λ to $\lambda w - \lambda w(n/x)^{1/(1-\gamma)}$, the term $w - w(n/x)^{1/(1-\gamma)} + c$ will move up by proportionately less to $\lambda w - \lambda w(n/x)^{1/(1-\gamma)} + c$. This results in MU_c dropping by less than the factor λ. Looking over the above first-order condition, you can see that this will work to increase the marginal utility cost of an extra child when wages rise since MU_c is falling at a slower pace. Finally, if $c < 0$, then fertility increases with wages.

(2) *Fertility, n, is unambiguously rising in the productivity of the home sector, x.* The workings here can generate a baby boom. An increase in x lowers the opportunity cost of a kid as it reduces $w \times MC_n = wn^{1/(1-\gamma)-1}/[(1-\gamma)x^{1/(1-\gamma)}]$. It also implies that more children can be had without reducing the time spent at work and hence reducing consumption. In fact, by substituting the solution for n in the person's budget constraint, you can see that consumption actually remains constant, because x drops out of the resulting expression. Thus, $MU_c = \theta/(c + c) = \theta/[w - w(n/x)^{1/(1-\gamma)} + c]$ also remains constant.

4.1   The Baby Boom and Baby Bust

Return to figure 2.2. The secular decline in fertility can be explained by the relentless rise in real wages, w, over this time period. Real wages rose by a factor of 10. Real wages rose because technological progress in the market sector increased the marginal product of labor; that is, at any given level of employment, an extra unit of labor could produce more. This made labor more valuable. This increased the opportunity cost of having a child. The baby boom can be attributed to the reduction in the cost of raising children due to the invention of labor-saving household technologies associated with the Second Industrial Revolution. This is represented by a rise in x. Note, though, that rising wages and improvements in labor-saving household technologies affect fertility in opposite directions. So the pattern of fertility in figure 2.2 must be explained by the fact that a burst in technological progress in the household sector during the middle of the twentieth century temporarily outweighed the effect of rising wages.

4.2 A Diagrammatic Transliteration
Some diagrams can help ferret out the intuition underlying the above result.

4.2.1 The Consumption possibilities frontier
The consumption possibilities frontier facing the household is defined by

$$c + \mathfrak{c} = w - w \left(\frac{n}{x} \right)^{1/(1-\gamma)} + \mathfrak{c}. \tag{4.4}$$

This is just the budget constraint (4.2) with $\mathfrak{c}$ added to both sides. The consumption possibilities frontier gives the maximal combinations of $c + \mathfrak{c}$ and n that are available to the household. It is shown in figure 2.9 by the curve labeled PP. The curve slopes downward because its derivative is

$$\frac{d(c + \mathfrak{c})}{dn} = -\text{MRT} = -w \left(\frac{1}{x} \right)^{1/(1-\gamma)} n^{\gamma/(1-\gamma)}/(1 - \gamma) < 0. \tag{4.5}$$

The (absolute value of the) slope of the consumption possibilities frontier is called the *marginal rate of transformation*, MRT. It gives the rate at which consumption, $c + \mathfrak{c}$, must be sacrificed in order to produce an extra child, n. The consumption possibilities frontier is concave because

$$\frac{d^2(c + \mathfrak{c})}{dn^2} = -\frac{d\text{MRT}}{dn}$$

$$= -w \left(\frac{1}{x} \right)^{1/(1-\gamma)} n^{\gamma/(1-\gamma)-1} \gamma/(1 - \gamma)^2 < 0.$$

This concavity reflects that fact that children can be produced only at an increasing marginal cost. That is, as n rises, each additional child can be produced only at an increasing cost in terms of foregone consumption. The frontier hits the vertical axis at the point $c + \mathfrak{c} = w + \mathfrak{c}$, which can be seen by setting $n = 0$ in (4.4). Likewise, it touches the horizontal axis at $n = x(1 + \mathfrak{c}/w)^{(1-\gamma)}$, which can be deduced by setting $c + \mathfrak{c} = 0$ in (4.4).

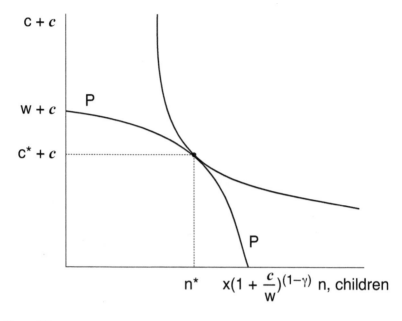

Figure 2.9
Determination of fertility, MRS = MRT.

4.2.2 Indifference curves over consumption and fertility The utility function (4.1) defines indifference curves over the various $(n, c + \mathfrak{c})$ combinations. The slope of an indifference curve gives the marginal rate of substitution, MRS, of children for consumption. By copying the steps used in section 2.2.1, it is easy to see that the slope of an indifference curve is

$$\frac{d(c + \mathfrak{c})}{dn}\bigg|_{\text{utility constant}} = -\text{MRS} = -\frac{MU_n}{MU_c} = -\frac{(1 - \theta)}{\theta}\frac{c + \mathfrak{c}}{n} < 0. \tag{4.6}$$

Or alternatively, in (2.3) just replace c with $c + \mathfrak{c}$ and l with n. Note that a fall in $c + \mathfrak{c}$ and a rise in n both lead to an indifference curve becoming flatter, implying that it is convex. More formally,

$$\frac{d^2(c + \mathfrak{c})}{dn^2}\bigg|_{\text{utility constant}} = \left[\frac{(1 - \theta)}{\theta}\right]^2 \frac{c + \mathfrak{c}}{n^2} + \frac{(1 - \theta)}{\theta}\frac{c + \mathfrak{c}}{n^2} > 0,$$

where equation (4.6) has been used for $d(c + \mathfrak{c})/dn|_{\text{utility constant}}$.

4.2.3 The determination of fertility The equilibrium level of fertility and market consumption is shown in standard fashion by the point $(n^*, c^* + \mathfrak{c})$ where the indifference curve is tangent to the consumption possibilities

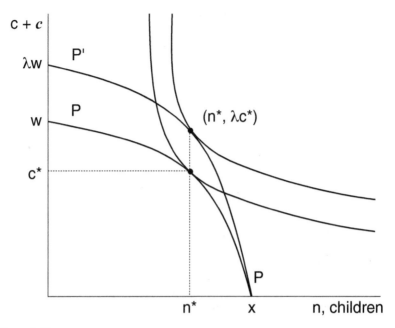

Figure 2.10
Effect of an increase in wages on fertility when $\mathfrak{c}=0$.

frontier (see figure 2.9). At the point $(n^*, c^* + \mathfrak{c})$, the marginal rate of substitution, MRS, equals the marginal rate of transformation, MRT.

(1) Case $\mathfrak{c}=0$. Let wages increase by a factor of λ and assume that $\mathfrak{c}=0$. In response, the consumption possibilities frontier will rotate upward from the curve PP, by a factor of λ, to the position shown by the curve $P'P$ (see figure 2.10). Thus, there is a positive income effect associated with an increase in wages. The slope of the consumption possibilities curve will increase by a factor of λ at any n point, too, as is evident from (4.5). That is, the marginal cost of an extra child rises. This effect should operate to reduce fertility. It's easy to deduce that consumption, c, will move up by a factor of λ and that fertility, n, will remain constant. This transpires because the substitution and income effects on fertility from an increase in wages exactly cancel out—a property of the logarithmic form of preferences adopted in (4.1). To see this, note that along any vertical line the slopes of the indifference curves increase in proportion with the increases in c, as is clear from (4.6). The slope of the indifference curve at the point $(n^*, \lambda c^*)$ is higher by a factor of exactly λ relative to the slope of the indifference curve at the point (n^*, c^*).

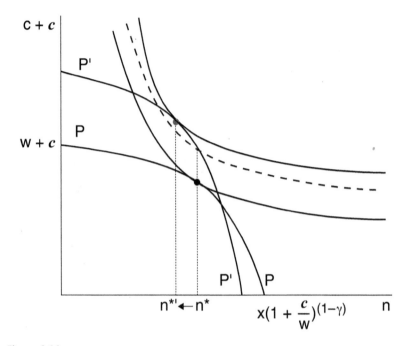

Figure 2.11
Effect of an increase in wages on fertility when $c > 0$.

(2) Case $c > 0$. Now suppose that wages jump up by a factor of λ, and assume that $c > 0$. The consumption possibilities frontier no longer shifts upward in a proportional manner. The horizontal intercept now shifts (see figure 2.11). A higher wage rate implies that the household production of goods, c, now frees up less time for kids. As can be seen, fertility must unambiguously fall from n^* to $n^{*'}$. Why? Suppose that fertility remains fixed at its old level, n^*. The slope of the consumption possibilities frontier will once again increase by a factor of λ, in line with (4.5); that is, it will increase from MRT to MRT$' = \lambda$ MRT. The slope of the indifference curve on the new production possibilities frontier at the point n^* will increase by less, though, due to the presence of the $c > 0$ term in preferences (see figure 2.11, where the dashed indifference curve hits the P'P' curve at the initial level of fertility, n^*). The slope of the indifference curve at this point will be given by MRS$' = [(1-\theta)/\theta](c+c)'/n^* = [(1-\theta)/\theta][\lambda w - \lambda w(\frac{n^*}{x})^{1/(1-\gamma)} + c]/n^*$, where (4.4) has been used. Let MRS represent the slope of the indifference curve at the old point. Now, MRS$' < \lambda$MRS, because $[(1-\theta)/\theta][\lambda w - \lambda w(\frac{n^*}{x})^{1/(1-\gamma)} +$

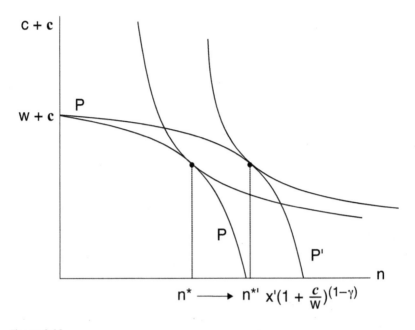

Figure 2.12
Effect of an improvement in household technology on fertility.

$c]/n^* < \lambda \times [(1-\theta)/\theta][w - w(\frac{n^*}{x})^{1/(1-\gamma)} + c]/n^*$. Hence, a point of tangency cannot occur.

What does this mean? At the point in question, a parent is willing to forgo having a child for MRS' units of consumption. According to their consumption possibilities, they can get MRT' units of consumption for an incremental cut in fertility. Now, MRS' < MRT'. Therefore, they should cut their level of fertility, because the consumption cost of an extra child, as measured by their MRT', exceeds what they are willing to pay, as reflected by their MRS'. In other words, when $c > 0$, the substitution effect from an increase in w outweighs the income effect. (It is easy to check that fertility will rise with wages in the case where $c < 0$.)

Last, consider the outcome of technological progress in the household sector. An increase in x shifts the consumption possibilities frontier outwards in the manner shown by figure 2.12 (from PP to PP'). At any point for n, the consumption possibilities curve becomes less steep since the consumption cost of an extra child falls. As a result, both the income and substitution effects operate to increase fertility. (Since kids are a normal good, as one moves upwards along any vertical line the slopes of the indifference curves increase—again, see (4.6). This implies that the new consumption point must lie to the right of n^*.)

4.3 Population Dynamics

How does fertility today relate to the size of the population tomorrow? The answer is simple. In the current framework an adult lives for only one period. Denote today's size of the adult population by p and let the size of the adult population tomorrow be represented by p'. Then, in the current set-up

$$p' = np.$$

This equation is easy to explain. Currently, there are p parents, and each parent has n kids. Therefore, in next period there will be np adults, given that the current stock of adults will have died.

The population is in a steady state when $p' = p$ across all adjacent time periods. Here there is no change in the population over time. This can happen only when $n = 1$, or when exactly one child is born for each adult. (When there are two parents in a family, then there must be two children to have a stable population. The ratio of children to adults is still 1:1.) Clearly the population will grow when fertility per adult, n, exceeds 1. Likewise, it will decline when fertility per adult is less than 1. Thus,

$$p' \gtreqless p, \text{ as } n \gtreqless 1.$$

The *gross* rate of growth in the population is given by p'/p. The *net* rate is $p'/p - 1$. Whether the above condition, specifying when the population will grow, remains constant or falls will play a key role in section 9 on the Malthus model.

5   The Size and Start of the Baby Boom in OECD Countries

Suppose that technological progress in the household sector around World War II lowered the cost of having children. If this burst in technological progress temporarily outweighed the effect of rising wages, one would expect to see a baby boom around this time. Households living in richer countries should on average have been better able to afford labor-saving household goods. A question arises: Was the baby boom bigger in richer countries? To answer this question, data on income and fertility is collected

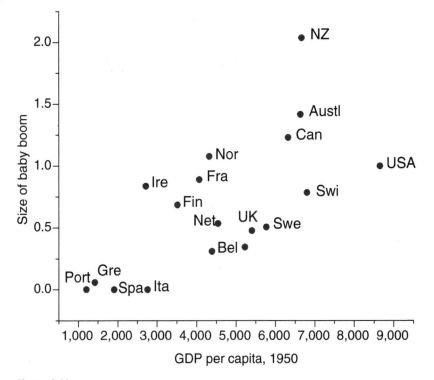

Figure 2.13
Cross-country relationship between the size of the baby boom and income.
Source: Greenwood, Seshadri, and Vandenbroucke 2005.

for 18 OECD countries—all those for which data is available. For each OECD country a graph similar to figures 2.6 to 2.8 is constructed. The baby boom is measured by the area below the fertility curve and above the horizontal line connecting the dates for the beginning and the end of the baby boom, as shown in figure 2.8 for Switzerland. The income data comes from the *Penn World Table 5.6*, discussed in Heston and Summers (1991), and measures a country's real GDP in 1950.

The results of this exercise are plotted in figure 2.13. As can be seen, there is a positive relationship between the size of the baby boom and a country's income. The Pearson correlation coefficient, a measure of the linear association between two series, has a value of 0.68. (See the Mathematical Appendix for more detail). There is no reason to presume that the relationship between the two variables is linear. Kendall's τ gives a nonparametric

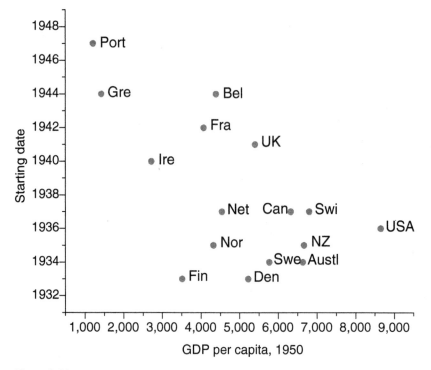

Figure 2.14
Cross-country relationship between the start of the baby boom and income.
Source: Greenwood, Seshadri, and Vandenbroucke 2005.

measure of the association between two series (again, see the Mathematical Appendix). A value of 0.48 is obtained for Kendall's τ. By either measure the two series are positively related to one another.

Likewise, one would expect that the baby boom should have started earlier in richer countries. This appears to be true. Figure 2.14 shows that there is a negative relationship between the start of the baby boom and a country's income. The Pearson correlation coefficient between the two series is −0.62. Similarly, the Kendall rank correlation coefficient is −0.31. Finally, for a *very limited* set of countries it is possible to plot the relationship between the size and start of the baby boom, on the one hand, and a measure of modern household technology adoption, on the other. For each of the six countries reported, a simple average is taken across the diffusion rates (which measure the percentage of households that have adopted a good) for hot running water, washing machines, refrigeration, sewing

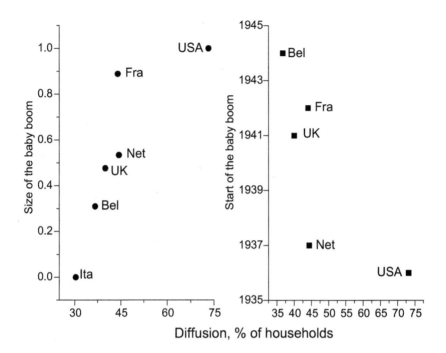

Figure 2.15
Cross-country relationship between the size/start of the baby boom and the diffusion of household technologies.
Source: Greenwood, Seshadri, and Vandenbroucke 2005.

machines, stoves (except the United States), and automobiles. Figure 2.15 shows the scatter plots. As can be seen, the size of the baby boom appears to be positively correlated with the diffusion of household technologies, while the start of the baby boom is negatively associated with it. This is what the theory suggests.

6   Advances in Obstetric and Pediatric Medicine

There were also considerable advances in obstetric and pediatric medicines, which significantly lowered the expected cost for a woman of having children. In 1915 there were 690 maternal deaths for every 100,000 live births, as is shown in figure 2.16. This dropped to 7.1 by 1995, a 100-fold decline. In the early days the main sources of these deaths were sepsis, toxemia, obstructed labor, and hemorrhages. Sepsis results from a bacteria in the

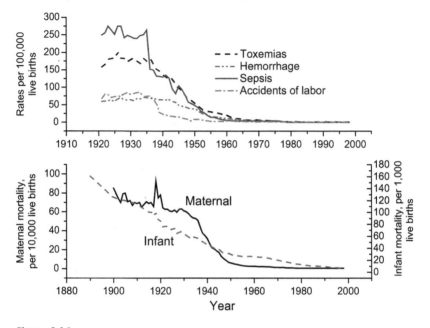

Figure 2.16
Deaths associated with childbearing.
Source: Albanesi and Olivetti 2014.

blood that can cause life-threatening infections. Toxemia (pre-eclampsia) is a high blood pressure disorder that can lead to fatal convulsions and seizures in a pregnant woman. The cause is still unknown, and the only treatment is the delivery of the baby. These maladies were greatly reduced due to the introduction of sulfonamide drugs (the mid-1930s), penicillin (early 1940s), the availability of blood transfusions due to blood banks (mid-1930s) and the increased use of hospitals for childbirths. Penicillin and sulfonamide drugs could be used to treat the bacterial infections arising from child-birth, such as sepsis. Blood transfusions allowed the blood lost due to hemorrhages to be replaced. The percentage of births that occurred in hospitals rose precipitously from 37 percent in 1935 to 82 percent in 1942, then to 94 percent by 1955. Advances in obstetric and pediatric medicine greatly reduced both infant and maternal mortality, as can be seen in figure 2.16. Associated with the decline in infant and maternal mortality was a rise in the general health of women who had given birth and of their babies. This greatly reduced the time spent with illnesses and disabilities.

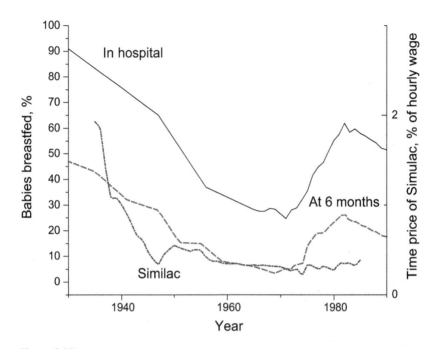

Figure 2.17
Breastfeeding and the time price of infant baby formula.
Source: Albanesi and Olivetti 2014.

Most babies were breastfed until the beginning of the last century. This tethered a mother to her child. Around 1920 a water-based infant formula was invented by Alfred W. Bosworth, a chemist and pediatrician, which mimicked the content of fat, proteins, and carbohydrates in maternal milk. (In the early part of the twentieth century mothers could use inferior cow-milk modifiers that were distributed by pediatricians.) Upon commercialization of this formula, breastfeeding dropped precipitously, as shown in figure 2.17. This was encouraged by the drop in the price of Similac, the first commercially available infant formula. Now, third parties (other family members or child care providers) could look after babies, which lowered the cost of having children for a mother. In recent times there has been a return to breastfeeding, as evidence on its health benefits for children has accumulated. Breast pumps have made this easier for mothers. While Orwell H. Needham filed a patent for a breast pump in 1854, they have been widely available only since the 1990s. At that time, Medela introduced its first electric, vacuum-suctioned breast pump.

6.1   The Model

The model developed in section 4 will now be extended to study the impact that advances in obstetric and pediatric medicines have on fertility. Suppose that a prospective mother knows that she will be sick with probability σ as the result of a pregnancy. She will remain healthy with probability $1 - \sigma$. If she becomes sick, then her productivity in child rearing will be $x = \lambda$. If she is healthy, then it will be $x = \eta$, where $\lambda < \eta$. Therefore, a healthy mother is more productive in child rearing. The woman must pick the number of her children before she knows the status of her health as a mother. So, with probability σ her utility will be $\theta \ln[w - w(n/\lambda)^{1/(1-\gamma)} + c] + (1 - \theta) \ln n$, and with probability $1 - \sigma$ it is $\theta \ln[w - w(n/\eta)^{1/(1-\gamma)} + c] + (1 - \theta) \ln n$. Utility in the first case is lower than in the second, for a given n.

Now, the woman desires to maximize the expected value of her utility. This is just the sum of the utilities for the two events, sick and healthy, where each utility is multiplied by the odds of its occurrence; i.e., it is simply the average of the utilities across the two events. Thus, her *expected* utility maximization problem is

$$\max_{n} \{\sigma\theta \ln[w - w(n/\lambda)^{1/(1-\gamma)} + c] + \sigma(1 - \theta) \ln n$$

$$+ (1 - \sigma)\theta \ln[w - w(n/\eta)^{1/(1-\gamma)} + c] + (1 - \sigma)(1 - \theta) \ln n\}.$$

This simplifies to

$$\max_{n} \{\sigma\theta \ln[w - w(n/\lambda)^{1/(1-\gamma)} + c]$$

$$+ (1 - \sigma)\theta \ln[w - w(n/\eta)^{1/(1-\gamma)} + c] + (1 - \theta) \ln n\}.$$

The first-order condition connected with this problem is

$$\frac{\sigma\theta}{w - w(n/\lambda)^{1/(1-\gamma)} + c} \times w \times \frac{1}{(1 - \gamma)} \frac{n^{1/(1-\gamma)-1}}{\lambda^{1/(1-\gamma)}} + \frac{(1 - \sigma)\theta}{w - w(n/\eta)^{1/(1-\gamma)} + c}$$

$$\times w \times \frac{1}{(1 - \gamma)} \frac{n^{1/(1-\gamma)-1}}{\eta^{1/(1-\gamma)}} = \frac{(1 - \theta)}{n}.$$

The left-hand side gives the *expected* marginal cost, $E[MC(\sigma)]$, of having an extra child, while the right-hand one represents the marginal benefit, MB. The expected marginal cost of a child is just an average of the marginal

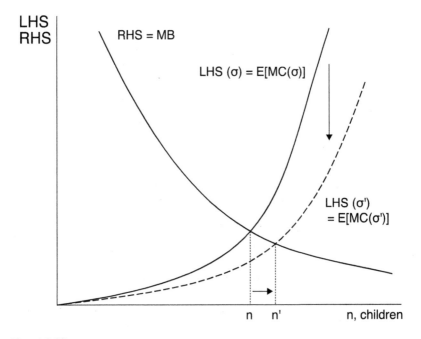

Figure 2.18
Impact of improvements in obstetric and pediatric medicines on fertility.

costs across the sick and healthy events. The marginal costs for each of these events can be broken down in exactly the same way as was done for the left-hand side of equation (4.3) in the earlier analysis of fertility. Clearly, the expected marginal cost is a function of the probability of becoming sick as a result of pregnancy, σ. The expected marginal cost is rising in the number of children, n. This is shown by the curve marked LHS (for left-hand side) in figure 2.18. The marginal benefit is decreasing in n and is portrayed by the locus labeled RHS (right-hand side). There is no uncertainty about the marginal benefit of a child, since this is assumed not to be affected by the mother's health status.

Suppose that advances in obstetric and pediatric medicines reduce the odds of a woman becoming sick; that is, assume that σ falls to σ'. This leads to the LHS curve moving downward. This transpires because

$$\frac{1}{w-w(n/\lambda)^{1/(1-\gamma)}+c} \times \frac{1}{\lambda^{1/(1-\gamma)}} > \frac{1}{w-w(n/\eta)^{1/(1-\gamma)}+c} \times \frac{1}{\eta^{1/(1-\gamma)}},$$

when $\lambda < \eta$. This inequality implies that the marginal cost of the extra child is higher when the woman becomes sick from her pregnancy than when she doesn't. So, when σ falls in the left-hand side of the above first-order condition, more weight is now being applied to smaller term. Hence, a reduction in σ, and a rise in $1 - \sigma$, will cause the LHS side curve to move down. Fertility therefore rises because the expected marginal cost of having an extra child has fallen.

7   Fertility and Wars: The Case of World War I in France

World War I was a cataclysmic event, which lasted four years, from 1914 to 1918. At the time France had a population of 40 million. There were 1.4 million military casualties in France. Additionally, around 300,000 civilians died either from the fighting or indirectly through starvation. Another 4.2 million soldiers were wounded. The deaths amounted to 4 percent of the population and casualties of over 10 percent.

Fertility plunged in Europe during World War I. This is shown in figure 2.19, which plots the crude birthrates for Belgium, France, Germany, Italy, and the United Kingdom. The crude birthrate fell by about 50 percent in each of these countries. For France this drop amounted to an estimated 1.4 million children not being born, or 3.5 percent of the French population at the time. This figure is comparable to the military losses from the war. The resulting shortfall in the French population was felt for many years, as figure 2.20 illustrates. The diagram portrays the demographic structure of the French population in 1950. It shows the contribution of each birth cohort to the French population, which is measured by the shaded area. As can be seen, there is a sizable dent in the population structure arising from the 1914–1918 birth cohorts.

The usual explanation for the drop in fertility is the massive mobilization of men required for fighting the war. A total of 8.5 million men served in the French army over the course of the war. The size of the age 20–50 male population is estimated to have been 8.7 million on January 1, 1914. Thus, almost all men served at some point during the war. This explanation has been challenged recently by Vandenbroucke (2014). He notes that somewhere between 30 and 50 percent of the men who were mobilized served away from the front and had opportunity to see their families. Furthermore, starting in 1915, the men at the front were granted regular leaves and could

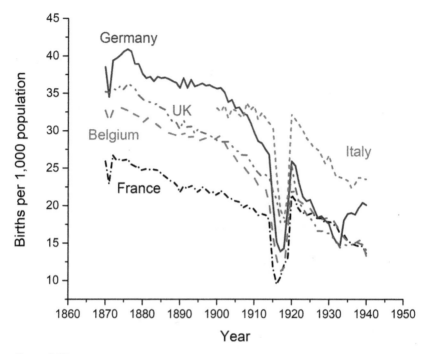

Figure 2.19
Drop in fertility for several European countries during World War I.
Source: Vandenbroucke 2014.

go home. Also, note that fertility jumped up in France immediately following World War I, as is evident from figure 2.19. Clearly, there were fewer
men alive then. So, by the missing-men logic, fertility should have fallen.
A commonsense explanation for the decline in fertility is that World War I
was just not a good time to start a family. A wife would face the prospect
that her husband could be maimed or killed in battle. Can this uncertainty
about family income lead to the observed decline in French fertility? Yes, is
the answer.

7.1 The Model

A simple adaptation of the framework used in section 4 suffices for the
problem at hand. Let tastes be given by the quadratic utility function

$$\alpha c - \frac{\beta}{2}c^2 + \gamma n - \frac{\chi}{2}n^2, \text{ with } \alpha, \beta, \gamma, \chi > 0, \text{ and for } c < \alpha/\beta, n < \gamma/\chi,$$

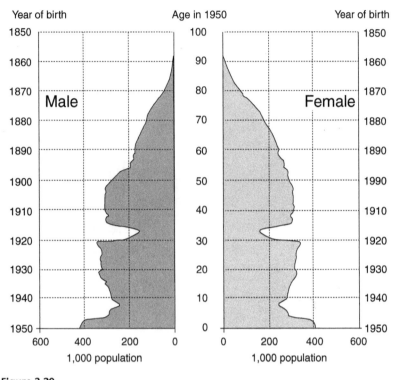

Figure 2.20
French age pyramid in 1950.
Source: Vandenbroucke 2014.

where again c and n are consumption and fertility. Recall from chapter 1 that a quadratic utility function is defined only for the range of its arguments that ensure utility is increasing in c and n. Suppose that there is a man and woman in the household. The man has a fixed workweek, which is normalized to 1. He earns the wage rate w. The woman divides her time between working and raising children. The woman's wage rate is ϕw. Children are costly in terms of the woman's time. Write a production function for children of the form

$$n = \frac{1}{\theta} l.$$

Now, assume that the man in the household dies with probably δ. In this case, the woman will have to raise any children by herself. So, with probability $1 - \delta$, the husband survives the war, and the household will earn $w + \phi w - \theta \phi w n$. With probability δ, he dies, and the household will have to survive on the woman's income alone, which will be $\phi w - \theta \phi w n$.

The household desires to maximize the expected value of its utility. Its optimization problem is

$$\max_{n} \{(1 - \delta)\alpha[w + \phi w - \theta \phi w n] - (1 - \delta)\frac{\beta}{2}[w + \phi w - \theta \phi w n]^2$$

$$+ \delta \alpha[\phi w - \theta \phi w n] - \delta \frac{\beta}{2}[\phi w - \theta \phi w n]^2$$

$$+ \gamma n - \frac{\chi}{2}n^2\}.$$

The first line gives the utility associated with consumption when the husband survives the war. This is weighted by the odds of survival, $1 - \delta$. The second line gives the utility from consumption when he dies. This event occurs with probability δ. The term on the third line is the utility arising from children. Since this is picked before the couple knows the outcome of the war, they know this term with certainty. The above maximization problem can be reformulated as

$$\max_{n} \left\{ \alpha[(1 - \delta)w + \phi w - \theta \phi w n] - (1 - \delta)\frac{\beta}{2}[w + \phi w - \theta \phi w n]^2 \right.$$

$$\left. - \delta \frac{\beta}{2}[\phi w - \theta \phi w n]^2 + \gamma n - \frac{\chi}{2}n^2 \right\}.$$

The first-order condition associated with this maximization problem, assuming an interior solution, is

$$\underbrace{\{\alpha - (1 - \delta)\beta[w + \phi w - \theta \phi w n] - \delta \beta[\phi w - \theta \phi w n]\} \times \theta \phi w}_{E[MC(\delta)]} = \underbrace{\gamma - \chi n.}_{MB}$$

The first-order condition is linear in n. This is a feature of optimization problems with a quadratic objective function and linear constraints. They are called *linear-quadratic optimization problems*. The left-hand side of the first-order condition is the expected marginal cost of having children, $E[MC(\delta)]$, while the right-hand side is the marginal benefit, MB. Since the expected marginal cost of an child is just an average of the marginal costs across the two events of the husband living or dying in the war, it is a function of the probability of him dying in the war, δ. It is easy to see that the expected marginal cost, $E[MC(\delta)]$, is increasing in δ, because $w + \phi w - \theta \phi w n > \phi w - \theta \phi w n$. Now, $w + \phi w - \theta \phi w n$ is the household's consumption when the husband survives the war, and $\phi w - \theta \phi w n$ is its consumption when he doesn't. The marginal utility of consumption is given by $MU_c = \alpha - \beta c$. Therefore, the marginal utility of consumption is higher in the situation in which the husband dies. Resources are scarce then, so the utility cost of an extra child resulting from the foregone consumption will be high. Therefore, the presence of war raises the expected cost of an additional child.

It is assumed that the marginal benefit of having a child is not a function of whether or not the husband survives the war. Consequently, there is no uncertainty here. The interior solution to the above linear first-order condition is given by

$$
\begin{aligned}
n &= \frac{-\alpha \theta \phi w + (1 - \delta)\beta(w + \phi w)\theta \phi w + \delta \beta \phi w \theta \phi w + \gamma}{\beta(\theta \phi w)^2 + \chi} \\
&= \frac{-\alpha \theta \phi w + \beta(w + \phi w)\theta \phi w - \delta \beta w \theta \phi w + \gamma}{\beta(\theta \phi w)^2 + \chi}.
\end{aligned}
$$

It's clear, therefore, that an increase in the probability of a husband not returning from the war, or δ, will reduce fertility, as expected.

8   The Choice between Jobs and Kids

The fact that many families desire to have children may influence the types of education and jobs that women take. There has been considerable convergence in the levels of educational attainment by men and women. In 2010–2011 a higher share of bachelor (57.2 percent), masters (60.1 percent), and Ph.D. degrees (51.4 percent) was awarded to women. There remained significant differences in fields of specialization. Out of bachelor degrees conferred then, the fraction going to women was only 17.6 percent in computer science, 17.2 percent in engineering, and 29.8 percent in economics, whereas women's share was 79.6 percent in education, 85.0 percent in health, and 77.0 percent in psychology.

College-educated women are much less likely to work in business and science-related occupations than are college-educated men, even when their college major was in these fields. This can be seen in table 2.1. Business and science-related occupations tend to pay more than other occupations, so this implies that women on average will be earning less than men. As discussed in section 7 of chapter 1, this will have implications for the mea-

Table 2.1
Share working in business/science occupations by age 30–35, 2000.

College Major	Men	Women
Business/Science	0.808	0.574
Humanities/Other	0.415	0.204

Source: Bronson 2015, table 5.

Table 2.2
Wage penalties for a woman taking time off.

	Bus/Sci occ		Other occ	
	Bus/Sci major	Other major	Bus/Sci major	Other major
Penalty	21.2%	15.9%	3.1%	2.8%

Source: Bronson 2015, table 6.

sured gender wage gap. Why would women choose work in lower paying occupations? The reason might have something to do with the necessity of taking time off to bear and raise children. Taking time off in business and science-related occupations is much more costly for a woman in terms of lost wages than taking time off in other occupations.[3] Table 2.2 shows the loss in wages for science and business occupations versus the loss for other jobs.

8.1   The Framework for Analysis

Consider the following model of fertility and job choice in a unisex household. The framework emphasizes the impact of labor supply-side considerations on the gender wage gap, induced by the desire to take certain jobs due to the fact that they facilitate having children. This is in contrast to the model presented in section 8 of chapter 1, which stresses the fact that the gender wage gap could partially arise from demand side considerations resulting from the fact that different jobs require different inputs, such as brawn and brain. Discrimination could also account for part of the gender wage gap, as discussed in section 1 of chapter 4.

3. Taking time off is defined as leaving the workforce for more than 9 months in the past 2 years.

Tastes are given by

$c + \theta \ln n$, for $\theta \in [0, \bar{\theta}]$,

where c is consumption and n is the number of kids the household has. A household is indexed by the (relative) weight it puts on the value of time at home, θ. Assume that $\theta \in [0, \bar{\theta}]$ is distributed across households according to a uniform distribution, $\theta/\bar{\theta}$. (See the Mathematical Appendix for the definition of the uniform distribution.) Thus, the fraction of households that have a weight, $\tilde{\theta}$, less than or equal to θ is given by

$\Pr[\tilde{\theta} \le \theta] = U(\theta) \equiv \theta/\bar{\theta}$.

As θ rises from its lower bound, 0, to its upper bound, $\bar{\theta}$, this fraction moves upward monotonically from 0 to 1. The household can choose between two types of jobs. The first is a high-paying job that pays a wage rate of v when the person works one unit of time on the job. The second is a low-paying job that pays a wage rate of w. To ensure that consumption is always positive the following condition is imposed.

Condition 1 (Positive Consumption) $w > \bar{\theta}$.

The *effective* time cost of having n kids for a person working in the high-paying job is

$l = \eta n$,

while the cost for a person employed in the low-paying job is

$l = \lambda n$,

where $\eta > \lambda$. (Think about effective time as measuring the amount of productive labor effort that the person can devote to their job.) Thus, relative to the low-paying job, having children while working in the high-paying job reduces the effective amount of time that a person can spend working in the market. The person is free to choose which job they would like.

8.2 Picking between Jobs and Kids

The choice between jobs and kids is broken down into two stages. First, attention is directed to the number of children a person will have, contingent on the choice of a job. Second, the focus shifts to the selection of a job, assuming that the individual will pick fertility optimally given the job choice.

Let v and w increase equiproportionately. Average fertility falls for two reasons. First, hold θ^* fixed in the above expression. As v and w move up, average fertility will drop. This occurs because both types of households will desire fewer children. Second, as w and v rise, θ^* moves up. As discussed, this will also lead to a fall in average fertility, as more households take the high-paying job, which has a lower level of fertility.

8.3 The Gender Gap

In the above framework, some households are taking lower paying jobs so that they may have more children. Children are costly in terms of time. The penalty for taking time off in a high-paying job is large. So, households that desire more children take a cut in the wage rate in the amount, $v - w$. The average wage rate in the economy is a weighted sum of the wages in high- and low-paying jobs, where the weights are the fractions of households taking each type of job. A fraction $U(\theta^*)$ of people will take the high-paying job, while $1 - U(\theta^*)$ will work in the low-paying one. Thus, the average wage is $U(\theta^*)v + [1 - U(\theta^*)]w$. The ratio of the average economy-wide wage to the wage in the high-paying job is

$$\phi = \frac{U(\theta^*)v + [1 - U(\theta^*)]w}{v}$$

$$= U(\theta^*) + [1 - U(\theta^*)]w/v.$$

Think about ϕ as reflecting the gender wage gap.

Now, if w and v increase equiproportionately, ϕ will rise or the gender wage gap will narrow. This occurs because θ^* gets larger. Why? As wages increase, the opportunity cost of having children rises. Both types of households will reduce their fertility. At the same time, more households will shift into the high-paying job. The high-paying job is more advantageous when households desire fewer kids.

9 Malthus

In October 1838, that is, fifteen months after I had begun my systematic inquiry, I happened to read for amusement Malthus on Population, and being well prepared to appreciate the struggle for existence which everywhere goes on from long-continued observation of the habits of animals and plants, it at once struck me that under these circumstances favourable variations would tend to be preserved, and unfavourable ones to be destroyed. The results of this would be the formation of a new species. Here, then I had at last got a theory by which to work.

—Charles Darwin (1876) *The Autobiography of Charles Darwin*

The English population showed little growth between 1200 and 1600. Thomas R. Malthus proposed a theory stating that the size of the population will be regulated by the productive capacity of the economy, which is limited by the availability of land. On the one hand, he felt that fertility increases with increases in income because parents can better support larger families. Likewise, the mortality rate declines with increasing income, since the diseases related to poverty fall. Thus, the population should rise with income. On the other hand, output per person should decline with population size due to diminishing returns to scale in production, which arise from the fixity of land. Thus, as the population expands, income will drop. These two considerations lead to a stable population size.

Figure 2.21 portrays the situation. Income per adult, y, is negatively related to size of the adult population, p, as the curve in the lower panel of the diagram portrays. This occurs because land is fixed in supply. Turn to the upper panel. Fertility is increasing with increasing income because parents can better support larger families. Likewise, the mortality rate declines with increasing income, since the diseases linked with poverty fall. The per-capita level of income associated with a stable population size, p^*, is given by y^*. Here the number of children being born (the fertility rate) is equal to the number of people dying (the mortality rate). If income was at some higher level, say y', then population size would increase. This would occur since fertility would exceed mortality. This expansion in population leads to a decline in income per person until it converges to y^*.

9.1 Setting up a Model

The basic model of fertility developed in section 4 will be used to model a Malthusian equilibrium. To this end, imagine an island, say England, inhabited by unisex households who have preferences defined over consumption, c, and kids, n, that are given by

$\theta \ln(c) + (1 - \theta) \ln n$, with $0 < \theta < 1$.

Write a production function for children in the form

$n = f^{1-\gamma}$, with $0 < \gamma < 1$,

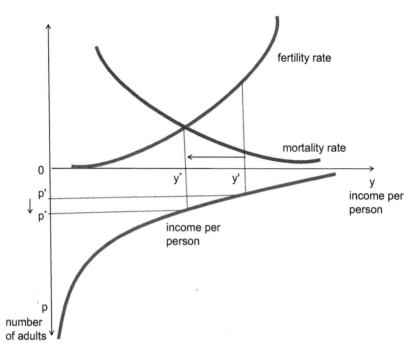

Figure 2.21
Malthusian equilibrium.

where n is the number of children and f is the food required to raise them. Additionally, there is one unit of land on the island. A unit of land produces ϕ units of consumption. The land on the island is split up equally among the adults living there. The current size of the adult population is p.

9.2 The Fertility Choice
Following the earlier analysis, it is easy to see that the household's maximization problem is

$$\max_{n}\{\theta \ln[\phi/p - n^{1/(1-\gamma)}] + (1 - \theta) \ln n\}.$$

Each household produces ϕ/p units of food from its plot of land. Out of this, $n^{1/(1-\gamma)}$ units of food must be used to feed the children. The first-order condition is

$$\underbrace{\frac{\theta}{\phi/p - n^{1/(1-\gamma)}} \times \frac{1}{(1-\gamma)} n^{1/(1-\gamma)-1}}_{MC = MU_c \times MU_n} = \underbrace{\frac{(1-\theta)}{n}}_{MB = MU_n}.$$

Once again, this just sets the marginal cost of a child, MC, equal to the marginal benefit, MB. The marginal cost of an extra child has two components. First, the extra child costs $MC_n = n^{1/(1-\gamma)-1}/(1-\gamma)$ in terms of additional food. Second, a one unit loss of food for the parent has a utility cost of $MU_c = \theta/[\phi/p - n^{1/(1-\gamma)}]$. The marginal benefit of an extra child is just $MU_n = (1-\theta)/n$, which by now doesn't need explaining. Solving for fertility, n, gives

$$n = \left(\frac{A}{1+A}\right)^{1-\gamma} (\phi/p)^{1-\gamma}, \tag{9.1}$$

whereas before $A \equiv (1-\gamma)(1-\theta)/\theta$. There are two features worth noting about this solution:

(1) Fertility, n, is increasing with increases in the productivity of land, ϕ. The more productive land is, the higher consumption will be, *ceteris paribus*. For any given level of n, ϕ lowers the marginal utility of consumption, $MU_c = \theta/[\phi/p - n^{1/(1-\gamma)}]$. A decline in the marginal utility of consumption leads the person to tilt expenditure toward having more children, because at the old level of fertility MC < MB.

(2) Fertility, n, is decreasing in the size of the population, p. The rationale for this is just the opposite of the first point.

9.3   Population Dynamics

Recall that the size of the population evolves according to the equation

$$p' = pn.$$

In the steady state of a Malthusian equilibrium the size of the population, p^*, does not change. This implies $p' = p = p^*$, so that n must equal 1. In other words, to have a stable population there must be exactly one child per adult. So, let $n = 1$. Then, it can be seen from (9.1) that

$$1 = \left(\frac{A}{1+A}\right)^{1-\gamma} (\phi/p^*)^{1-\gamma},$$

which implies that the steady-state size of the population, p^*, is

$$p^* = \frac{A}{1+A}\phi.$$

Clearly, an increase in the productivity of land, ϕ, will lead to a larger equilibrium population size, p^*. The steady-state level of fertility will remain constant at 1; population can remain constant only when there is one child for every adult.

Go back to equation (9.1), (i) this equation implies n is decreasing in p, and (ii) if $p = p^*$, then $n = 1$. Hence,

$$n \gtreqless 1, \text{ as } p \lesseqgtr p^*.$$

Suppose that p is below the steady-state size of the population p^*. Then, $n > 1$, from the above equation. Hence, the size of the population must grow according to the law of motion $p' = pn$. As the population increases, the food that each household produces, ϕ/p, decreases. This causes fertility, n, to fall, a fact evident from (9.1). As fertility drops, the gross rate of growth in the population, p'/p, will slow down, because $p'/p = n$. Eventually, it will converge to 1 (implying zero net growth) as p approaches p^*.

How does the above model relate to figure 2.21? Equation (9.1) defines the fertility locus $n = [A/(1 + A)]^{1-\gamma} y^{1-\gamma}$, where per-capita income is just $y = \phi/p$. Clearly, this locus is upward sloping (concave) and starts at the origin, as is shown in figure 2.22. All adults die at the end of each period so the mortality rate is 1.0. Thus, the mortality locus is just a horizontal line at 1.0. Using the relationship $y = \phi/p$, one can see that $p = \phi/y$. This locus is decreasing (and convex) in y. Now, suppose that land becomes more productive or that ϕ rises to $\phi' > \phi$. This change does not affect either the fertility or the mortality loci. It does, however, result in a downward shift in the productivity locus. Hence, population will rise to exactly the level that the new more productive land will sustain. Note that income per capita, y, remains unchanged.

9.4 Some Evidence

Turn now to figure 2.23, which plots the size of the English population between 1200 and 1860. Observe that there is a dip in the population around 1350. This is the Black Death, which spread to England from Europe in 1348. Historians have estimated that the plague killed somewhere between 25 to 65 percent of the English population. According to Malthus's hypothesis, the population should rebound to its old steady-state level following the plague. This is consistent with the data. Also, when the population is low, wages should be high. Think about this as being represented by per-capita income in the model, or ϕ/p. When p is small, then ϕ/p will be large. This also seems consistent with the data. As the

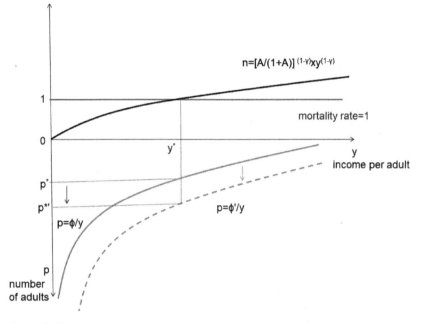

Figure 2.22
Impact of an increase in productivity from ϕ to ϕ' on Malthusian equilibrium.

population rebounded, wages fell. The Malthusian era ended sometime in the 1700s, when the British Industrial Revolution started. At that time both human and physical capital became more important in production relative to land. Unlike land, both human and physical capital can be augmented through education and investment.

10   Literature Review

The economics literature on fertility starts with Malthus (1798). The Malthusian theory of population determination is covered in section 9. A classic paper on fertility in modern macroeconomics is by Razin and Ben-Zion (1975), whose model is similar to the set-up presented in section 4. They develop a model of fertility in which kids simply enter their parents' utility function in the same way as other goods—say, as in (4.1). Becker and Barro (1988) construct another well-known model of fertility, which is more sophisticated, but harder to work with. Now parents care about the utility of their children in addition to the number of kids. Since a parent

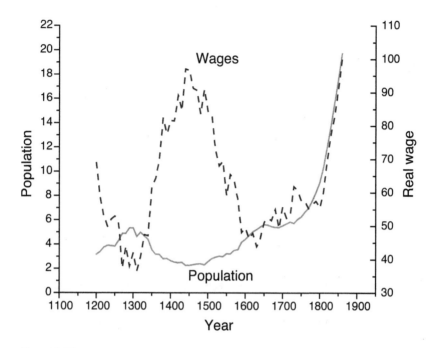

Figure 2.23
English population and real wages, 1200–1860.
Source: Clark 2010.

cares about the happiness of their child who will in turn care about the happiness of their child (and so on ad nauseam), the Becker and Barro (1988) model reformulates in terms of a person or dynasty who lives forever. Ermisch (2003) has a chapter on economic theories of fertility. Manuelli and Seshadri (2009) document and explain the negative correlation between fertility and income across countries. They build a model of fertility choice and investment in a child's human and health capitals. In a quantitative exercise, they find that differences in productivity across countries account for a large fraction of cross-country differences in fertility.

Over the epochs of European history, fertility has followed a ∩-shaped pattern. An important paper on the demographic transitions literature is by Galor and Weil (2000). They develop a model of the ∩-shaped pattern of fertility by combining elements of Malthus (1798) with Razin and Ben-Zion (1975). They also allow for parents to invest in the human capital of their children. In addition they make two key assumptions: first, technological progress is an increasing function of population size, and second, the return on education rises with the rate of technological advance. In their framework, the world rests in a Malthusian equilibrium for a long time. Per-capita income remains more or less constant, and all increases in aggregate income induced by technological progress are absorbed by expansions in the population. As the population slowly grows bigger, the pace of technological progress begins to pick up, and the economy exits the Malthusian regime. At first parents use the extra income generated by technological advance to have more kids, since the return on education is still low. As the rate of technological progress accelerates, the return to education rises, and parents choose to have fewer kids but they invest more in them.

The United States experienced only the downward part of the ∩. A model of the long-run decline in U.S. fertility in which the driving force is rising productivity, is developed by Greenwood and Seshadri (2002). They stress the movement in labor from agriculture to industry, where the latter industry is more skill-intensive than the former. The model includes an education decision. Their model mimics, quantitatively, the long-run U.S. fertility data and is also consistent with the secular rise in education and the reallocation of labor from agriculture to industry. Vandenbroucke (2008) models the populating of the United States in the nineteenth century, both through reproduction and immigration. He finds that technological progress, in the form of the transportation revolution (which lowered the cost of moving West), and wage growth had a larger impact on the geographic distribution of the population and its natural rate of increase than immigration into the United States.

A calibrated model that delivers a transition from Malthusian stagnation to growth, accompanied by a demographic transition from high to low fertility, is presented in Doepke (2004). The engine in his analysis is a Becker and Barro (1988)–style model modified to allow for parental human capital investment in children. He uses the model to study the role of social policies in shaping the demographic transition of a country. Another paper modeling the liftoff from a no-growth Malthusian regime to the modern era with growth is by Hansen and Prescott (2002). They stress the necessity to use land, which is fixed in supply, as the key factor explaining stagnation in the Malthusian era. In the modern era, production relies more heavily on capital, which is reproducible. The heavier reliance on reproducible capital allows for a higher growth rate. Hansen and Prescott (2002) present evidence showing how the value of farm land relative to GDP has fallen in the United States as production has shifted away from land-intensive agricultural methods.

As discussed, more than half of the drop in U.S. general fertility rate occurred before 1900. It would thus be hard to attribute the decrease in fertility to the introduction of modern contraceptives. Wilkinson (1973) discusses in detail how even primitive societies could regulate their fertility to keep them within the ecological limits of the land. Apparently, no society has had women being close to birthing the biologically maximum number of children. Eckstein, Mira, and Wolpin (1999) advance the hypothesis that a drop in infant mortality can explain the reduction in fertility. Like all hypotheses in economics, its validity has been questioned. In the United States the decline in fertility preceded the sustained drop in child mortality, which started in the 1890s. Fertility also dropped in France before child mortality fell. Theoretically speaking, under reasonable conditions on the utility function for children, a drop in infant mortality will lead to fewer pregnancies but more surviving children, see Greenwood, Guner, and Vandenbroucke (2017, proposition 11). Eckstein, Mira, and Wolpin (1999) look only at births, not surviving children. Doepke (2005) argues within the context of a sequential fertility model with uncertainty about child mortality that the impact of a decline in child mortality is likely to be small. That is, if the issue is child mortality, why would a woman who already has some surviving children give birth to yet more children, as is observed in the data? To see the idea in Doepke's research, suppose a family wanted 2 kids. Once they had 2 surviving offspring, then why would they continue having more children, say up to 5 or more, as did the average American woman who was born in 1800.

Another idea is that the fall in fertility resulted from the fact that children were no longer needed for either labor or old-age security. Historically, these factors would have defrayed the cost of children. A prohibition on child labor or the advent of old-age security would have increased the cost of kids. Hence, the mechanism at work is similar to rising wages. Doepke (2004) analyzes differences in the timing and pace of the demographic transition across countries (figure 2.1 shows some differences) and ascribes them to differences in child labor laws. Boldrin, de Nardi, and Jones (2015) argue that government-provided old-age pensions are strongly associated with low fertility. It is important to note that children were probably never capital goods that yielded a positive return. Sometime ago Mueller (1976) calculated that in peasant agriculture, "children—from birth to the time of their own marriages—tend to produce less than they consume."

Little work has been done on the underlying cause of the baby boom. The best known hypothesis is by Easterlin (1987). The generation that spawned the baby boom grew up during the hard times of the Great Depression and World War II. As a result, this generation had low material aspirations. They then entered the work force in the 1950s and 1960s, a good time economically speaking. Given their low material aspirations, they used family formation as an outlet for their earnings. This hypothesis is empirically flawed. The age of the women bearing children during the baby boom is hard to reconcile with a simple formulation of his theory.

Greenwood, Seshadri, and Vandenbroucke (2005) propose that the baby boom was the result of technological progress in the home, which economized on the need for labor. A complementary hypothesis is advanced by Albanesi and Olivetti (2014), who argue that advances in obstetric and pediatric medicines led to improvement in the health of new mothers and their children. All of the facts concerning these medical advances are taken from Albanesi and Olivetti (2014). Both hypotheses operate by reducing the time cost associated with having young children. Gershoni and Low (2017) discuss how the availability of in-vitro fertilization in Israel has extended women's fertility time horizons and has made it easier for them to pursue more advanced levels of education and attain better labor market outcomes. The analysis of fertility in France during World War I draws from Vandenbroucke (2014), who does not use the quadratic utility function relied on here, instead employing more standard isoelastic utility functions. A quadratic utility displays increasing absolute risk aversion, implying that the wealthy are less disposed to risk than the poor. The example in text is intended for illustrative purposes only.

Last, Bronson (2015) investigates the relationship between kids and the career choices made by women. She also evaluates different family-leave policies. Caucutt, Guner, and Knowles (2002) suggest that when the wage penalty associated with childbearing is high, women prefer to postpone their fertility and instead first build their human capital. Again, an elementary textbook on the economics of women in the labor force is by Blau,

Ferber, and Winkler (2014). The facts presented in section 6 on the fraction of university degrees awarded to women are taken from there. They also discuss family-leave policies.

11 Problems

(1) *Demographic transitions.* Consider a fertility model in which a parent's tastes are distributed over consumption, c, and kids, n, in the following manner:

$$\theta \frac{(c - \psi)^{1-\gamma} - 1}{1 - \gamma} + (1 - \theta) \ln n,$$

with $0 < \psi$, and $0 < \theta, \gamma \leq 1$. A single adult has one unit of time to split between working and having children. A unit of labor receives the wage rate, w, in the workplace. Assume that $w > \psi$. Each child costs α units of *time*.

(a) Formulate the maximization problem facing an adult.

(b) Derive the first-order condition determining the number of children, n. Explain the intuition underlying this first-order condition. (*Hint:* It may help to do the analysis in terms of two well-known curves in economics.)

(c) Suppose $\gamma = 1$. How does an increase in wages affect n? What is the intuition underlying this result? (*Hint:* Use the first-order condition found in point [b].) For the intuition it may help to note that as $\gamma \to 1$, it happens that $[(c - \psi)^{1-\gamma} - 1]/(1 - \gamma) \to \ln(c - \psi)$.]

(d) Suppose $\psi = 0$ and $\gamma < 1$. How does an increase in wages affect n? What is the intuition underlying this result?

(e) Suppose $\psi > 0$ and $\gamma < 1$. As wages grow from some very low level $[w < \psi(1 - \gamma)]$ to a very high level, what will the time path of fertility look like? (*Hint:* Think about a demographic transition.)

(2) *Calibrating a fertility model.* Take the model of fertility developed in the chapter, but change the production function for children to

$n = xl.$

Focus on four years in U.S. history, viz. 1800, 1940, 1960, and 2000. In these years fertility *per adult* was 3.5, 1.1, 1.8, and 1.0, respectively. Wages were 1, 5, 8, and 10. Assume that household productivity has the following time path: $x_{1800} = x_{1940} = 1$ and $x_{1960} = x_{2000} = ?$

(a) Start with 1800 and 1940. Are there values for c and θ so that the model will fit the data for these two years?

(b) Taking the above parameter values for c and θ as given, is there a value for x for 1960 so that the model fits this year?

(c) What is the model's prediction for the year 2000? Discuss.

3   The Decline in Marriage

The decline in marriage and the rise in positive assortative mating since World War II are the focus of this chapter. The analysis stresses the improvements in income and household technologies that have made it easier for a person to live alone. Positive assortative mating refers to the tendency of people to marry mates from the same socioeconomic class. In yesteryear, a woman's contribution to household production was a highly valued factor. But as this consideration dwindled, due in part to labor-saving household appliances and products, a woman's contribution to household income gained importance. This increased the incentive for a high-earning man to look for a high-earning woman, and vice versa. This tendency is amplified when there is a rise in income inequality caused by an increase in the ratio of skilled to unskilled wages.

The chapter begins by outlining two key motives for marriage—namely, love and money. It then lays out some facts regarding marriage, divorce, and positive assortative mating. To address the facts presented, a basic model of marriage is presented. The model stresses economies of scale in household consumption/production as an economic motive for marriage; that is, the cost of maintaining a two-person household is less than twice the cost of a one-person household. Three forces are stressed as driving the decline in marriage. First is the rise in wages, which makes establishing a one-person household more affordable. Second is labor-saving technological progress in the household sector, which implies less time needs to be spent maintaining a one-person household. The third is the rapid drop in the prices of time-saving goods used at home. These three forces reduced the importance of scale economies in the household consumption/production. The basic model of marriage is then extended to include divorce.

The issue of assortative mating is discussed next. To deal with this, an elementary version of the Gale-Shapley matching algorithm is introduced. Due to technological progress in the home, the value of women's household labor has declined. As more women entered the workplace, the importance of the earnings that they can bring home has increased. This section shows how a rise in market wages and technological progress in the home can lead to an increase in positive assortative mating.

Two side roads are then followed. The first is the condition of children who grow up living with a single parent, usually a mother. It is shown that, statistically speaking, these children have worse life outcomes in terms of education, employment, and teenage pregnancies than children who grow up living with two parents. In turn, children raised by single parents are more likely to become single parents themselves. A poverty cycle results. This is a very important social issue, and a simple model explaining the exhibited stylized facts is presented. The second takes up Becker's famous model of efficient marriages and assortative mating. As usual, the chapter concludes with a review of the literature upon which the chapter is based.

1   Love or Money

What determines whether or not a single person will marry? Likewise, when will a married couple divorce? Two motives for marriage are stressed here:

(1) *Love and Companionship.* "Come live in my heart and pay no rent," composed the Irish songwriter Samuel Lover (1797–1868). It is hard for an economist to improve on what a songwriter can say about love and companionship. The love between two people will be modeled here in a clinical fashion, as a term in tastes.

(2) *Economics.* Two people living together may have a higher level of material well-being than if they each live alone. This can happen because there are economies of scale both in the consumption of market goods and in the consumption/production of nonmarket ones. In time past the last factor was much more important than today. Two hundred years ago the United States was largely a rural economy. The household was the basic production unit, with the family producing a large fraction of what it consumed. At the time, most marriages were arranged by the parents of young adults. Key considerations were whether or not the potential groom would be a

good provider and the bride a good housekeeper. Ogburn and Nimkoff (1955, 40–41) quote *Godey's Lady's Book* (1831) which states "No sensible man ever thought a beautiful wife was worth as much as one that could make good pudding." It adds (1932), "Among our industrious fore-fathers it was a fixed maxim that a young lady should never be permitted to marry until she had spun for herself a set of body, bed and table linen. From this custom all unmarried women are called spinsters in legal proceedings." In modern times, living standards are much higher than in the past. This makes it easier to live alone. Additionally, much less time is spent in house-hold production, as discussed in chapter 1, section 1. Again, this makes it easier to live alone. With the rise in married female labor-force participa-tion, the earnings that a woman brings home becomes an important factor. Therefore, the income that each party can place on the table in a marriage will be a consideration in deciding whether to marry or not.

In modern times couples cohabitate without being married. Roughly 50 percent of 15 to 44 year olds have cohabited at some point in their lives. It has been reported that 50 percent of gays and 60 percent of lesbians live with their same-sex partners. Also, same-sex marriages have recently been legalized. Less is known about these forms of relationships. Historically data has been collected using the legal definition of marriage; in fact, these newer types of relationships were prohibited in various places at different points in time. Still, the economic considerations underlying the decision to marry developed here applies in large measure to these newer household types.

2 Definitions for Marriage and Divorce Rates

There are several concepts used to measure marriage and divorce rates within a society. Here are four:

(1) *Crude Marriage Rate*. This rate examines the number of marriages relative to the size of the population.

$$\text{CRUDE MARRIAGE RATE} = 1{,}000 \times \frac{\text{number of marriages in a year}}{\text{total population}}.$$

The multiplication by 1,000 results in the number of marriages per 1,000 people. This marriage rate is sensitive to the age structure of the population, because the marriage rate is much higher for the young than for the old (who are much more likely to be married already). So, for example, in 2000–2004 a 25-year-old never-married woman had about a 10 percent chance of marrying in the next year. This dropped to less than 2 percent for a 40-year-old woman.

(2) *Age-Specific Marriage Rate.* One can also measure the marriage rate for people in certain age ranges.

AGE-SPECIFIC MARRIAGE RATES (for people of age group i) =

$$= 1{,}000 \times \frac{\text{total number of people marrying in age group } i}{\text{population in age group } i}.$$

So, for example, one could look at the marriage rate for people aged 20 to 24 or 35 to 39. Countries with a younger population should have more marriages relative to countries with an older population. This rate controls for the age structure of the population.

(3) *Crude Divorce Rate.* This is the analogue of the crude marriage rate. It examines the number of divorces per 1,000 people.

$$\text{CRUDE DIVORCE RATE} = 1{,}000 \times \frac{\text{number of divorces in a year}}{\text{total population}}.$$

(4) *Age-Specific Divorce Rates.* This is the analogue of age-specific marriage rates. Specifically,

AGE-SPECIFIC DIVORCE RATES (for people of age group i)

$$= 1{,}000 \times \frac{\text{total number of people divorcing in age group } i}{\text{population in age group } i}.$$

Again, it controls for the age structure of the population.

3   Trends in Marriage and Divorce

3.1   Since World War II

To understand the importance of economic factors in determining marriage and divorce, consider some facts. A much smaller proportion of the adult population is now married compared with 50 years ago (figure 3.1, left-hand-side panel). In 1950, 75 percent of women were married (out of nonwidows age 15 or older). By 2016, this had declined to 56 percent. Note that for every married woman there is a married man and, vice versa— ignoring recently legalized same-sex marriages. Adults now spend a smaller fraction of their lives married. In 1950 women spent about 88 percent of

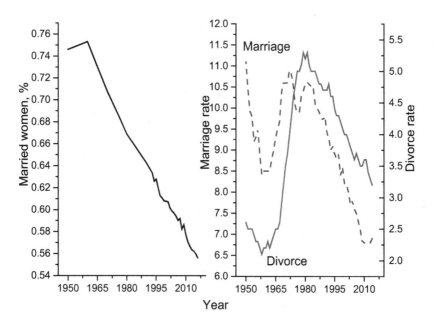

Figure 3.1
Marriage and divorce, 1950–2016. The left-hand-side panel shows the percentage of
nonwidowed women (ages 15 and above) who were married. The right-hand-side
panel plots the crude marriage and divorce rates.
Source: Greenwood, Guner, and Vandenbroucke 2017.

their life married as compared with 60 percent in 1995. Underlying these
trends are two factors:

(1) Between 1950 and 2016, the crude marriage rate declined drastically. In
1950 there were 11 marriages per 1,000 people, as compared with just 7 in
2016.

(2) Between 1950 and 2016, the crude divorce rate spiked and then declined
(figure 3.1, right-hand-side panel). Nonetheless, today the crude divorce
rate of 3.2 per 1,000 people exceeds the 1950 value of 2.6.

Why do these facts illustrate the importance of economic considerations
in determining marriage and divorce? Ogburn and Nimkoff (1955) felt
that if the influence of economic factors dwindled over time, then more
weight would be placed on love and companionship as the motive for
marriage. Economic factors have declined in importance due to both rising living standards and time-saving household products. Rising living

standards imply that over time it becomes more affordable for people to live alone. Likewise, labor-saving household products mean that it is much easier for a single person to undertake household production. The upshot of this is that people become pickier about their mate. This leads to fewer marriages and more divorces.

3.2 The Entire Twentieth Century

The story appears to be more nuanced if one goes back farther in time. Figure 3.2 plots the percentage of women (nonwidowed between the ages of 18 and 64) who were married from 1880 to 2000. About 72 percent of the population was married in 1900, as opposed to 62 percent in 2000. Observe that the number of marriages shows a ∩-shaped pattern roughly coinciding with the baby boom years. This pattern is not as dramatic as it seems at first glance. The population was much younger at the turn of the last century than it is today. Women aged 18 to 30 made up 44 percent of the population in 1900. Now, they account for 26 percent. Young women are much less likely to be married than older ones. Figure 3.2 also shows the fraction of the female population that is married after making a correction for the shift in the age distribution. This correction assumes that the age structure of the population in the year 2000 holds for the entire time period.[1] First, note that a much higher percentage of women

1. The fraction of married women out of the total female population, f, can be written as

$$f = \sum_{i=1}^{I} \frac{m_i}{p_i} \frac{p_i}{p},$$

where m_i is the number of married women in age group i, p_i is the size of the age-i population, and p is the size of the total population. Observe that f depends both on the age composition of the population (or the p_i/p terms) and on the fraction of each age group who is married (or the m_i/p_i terms). Now, for any year t define an age-adjusted measure by

$$\widehat{f_t} = \sum_{i=1}^{I} \frac{m_{i,t}}{p_{i,t}} \frac{p_{i,2000}}{p_{2000}}.$$

Thus, $\widehat{f_t}$ calculates the fraction of women that would be married, if the age composition of 2000 was in effect at time t.

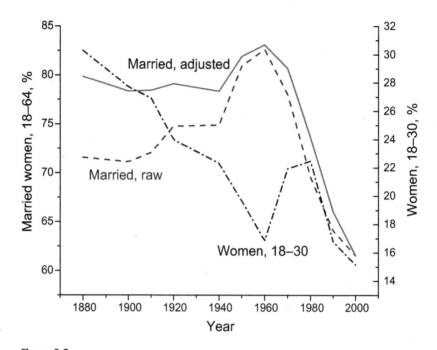

Figure 3.2
Marriage, 1880–2000.
Source: Greenwood and Guner 2009.

were married at the beginning of the century as opposed to the end, about 17 percentage points more. Second, the hump is still there, but it is much less pronounced.

At the beginning of the twentieth century the vast majority of never-married young women (close to 80 percent) lived as dependents with their parents. A substantial fraction lived in households as nonrelatives (i.e., as boarders, servants, and so on). Almost none lived in her own household, however. The fraction of young stand-alone households, with an independent female, made up by single women has become much more prevalent over time. It has risen from close to zero at the turn of the last century to about 50 percent today, as figure 3.3 illustrates. Additionally, figure 3.3 plots the proportion of young households made up by married couples. As can be seen, it fell from nearly 100 percent at the turn of the last century to less than 50 percent today. Interestingly, this plot shows a monotonic decline from roughly 1910 on; the hump has disappeared.

What can account for the hump-shaped pattern in marriage in figure 3.2? Specifically, why did the number of marriages rise between 1940
and 1960, and subsequently decline? The answer may be that labor-saving
technological progress in the household sector made it feasible to establish smaller and smaller households. In the initial stages of development,
technological advance made it easier for a young adult to leave their parent's
home and marry. As household technology progressed further, it became
viable for young adults to leave home and remain single. Therefore, the
move by young adults from large to two-person households coincided
with an increase in marriages, while the subsequent shift toward one-
person households was associated with a decline. This hypothesis is consistent with the decline in the ratio of young married households to
all young households with an independent female that is shown in
figure 3.3.

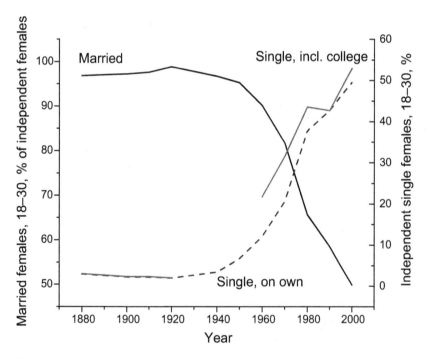

Figure 3.3
Living arrangements for young women, 1880–2000.
Source: Greenwood and Guner 2009.

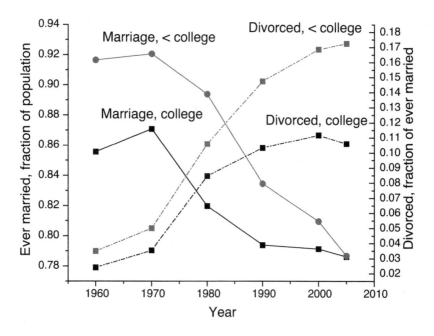

Figure 3.4
Decline in marriage and the rise in divorce by education level.
Source: Greenwood, Guner, Kocharkov, and Santos 2016.

3.3 Educational Attainment

Another interesting fact is that the decline in marriage and the rise in divorce are greater for non–college educated people than for college educated ones (see figure 3.4). This is exactly what the above hypothesis would predict. Non–college educated people are poorer. Therefore, in the past, they would have been more inclined to marry for economic reasons. As living standards rose and labor-saving household products became available, they should have experienced the biggest decline in marriage and the largest increase in divorce.

3.4 The Rise in Positive Assortative Mating

As more married women enter the labor force, their earnings in the workplace become a key economic factor in the decision to marry or not. A high-earning woman will desire a high-earning man, and vice versa, other things being equal. People at the lower end of the income distribution, while they might desire a richer mate, are likely to marry someone like

Table 3.1
Positive assortative mating, age 25–54.

	1960			2005	
Husband	Wife		Husband	Wife	
	< College	College		< College	College
< College	0.855 (0.821)	0.023 (0.056)	< College	0.545 (0.427)	0.108 (0.226)
College	0.082 (0.115)	0.041 (0.008)	College	0.109 (0.227)	0.237 (0.120)
$\chi^2 = 33,451$	$\rho = 0.41$	$n = 195,034$	$\chi^2 = 77,739$	$\rho = 0.52$	$n = 288,423$

Source: Greenwood, Guner, Kocharkov, and Santos 2016.

themselves due to competition in the marriage market. Hence, when individuals do marry, people are more likely today to pair with an individual from the same socioeconomic class than they were in yesteryear. The fact that people do not marry randomly is easy to illustrate using a contingency table.

Definition 1 (2 x 2 contingency table) *A 2 × 2 contingency table is an array with four entries that gives the observed frequencies of data that are classified according to two categories, each of which can take two values. The rows present one of the categories while the columns gives the other.*

Consider the contingency table shown in table 3.1. The rows give the husband's educational levels, the columns the wife's. First, when marrying, people tend to sort by educational class. The number in a cell shows the fraction of all matches that occur in the specified category. The number in parenthesis in a cell provides the fraction that would occur if matching occurred randomly.[2] Note that there is positive assortative mating. This occurs because the numbers along the diagonal for the data exceed those that would occur if the matches were random. The hypothesis

2. Let p represent the fraction of men with a less-than-college education and q denote the proportion of women who are less-than-college-educated. When matches are random, the contingency table appears as

Male	Female	
	< College	College
< College	pq	$p(1-q)$
College	$(1-p)q$	$(1-p)(1-q)$.

Observe that the entries sum to one.

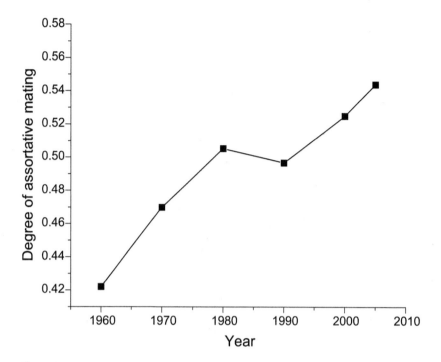

Figure 3.5
Positive assortative mating, as measured by the Pearson correlation coefficient. As calculated by Greenwood, Guner, Kocharkov, and Santos 2016.

of random matching is rejected by the χ^2 statistics. This is a statistical test that assesses whether the observed deviations from randomness could have occurred by chance; the extremely high value for the χ^2 test statistic says that this is very unlikely. Second, the extent of positive assortative mating has become stronger over time. This is shown by the Pearson's correlation coefficient, ρ, which measures the degree of association between the husband's and wife's educational categories. (see the Mathematical Appendix for more detail.) Figure 3.5 plots the rise in ρ over time.[3]

3. When calculating the Pearson correlation coefficient, a value of 1 is assigned when the person has a college degree and a value of 0 is given otherwise. So, each marriage consists of two numbers out of four possibilities, $(0,0), (0,1), (1,0)$, and $(1,1)$. The correlation measures the association between the first and second numbers in a marriage.

4   A Basic Model of Marriage

Suppose that each adult in a household has one unit of time that can be allocated to either working in the market or at home. Thus, a single household will have one unit of time while a married household has two. Let the market wage be represented by w, where it is assumed that a man and woman earn the same amount. Household production is done according to the home production function,

$$n = d^{1-\kappa} l^{\kappa}, \text{ with } 0 < \kappa < 1,$$

where l is the household's labor at home and d represents the inputs of goods into home production. The exponent κ reflects how important household labor is in home production. The larger it is, the more important is home work. Suppose that d is fixed and can be purchased at the price q. Let $q < w$ so that a single household can always afford to buy the inputs, d, needed to maintain a household. The fact that a married household has two units of time, as opposed to the single household's one unit, implies that the former will be able to better afford the inputs into home production. As will be seen, this introduces a scale economy into household consumption that favors married life over single life. For example, setting up a household for two people is less than twice the expense of setting up a household for a single person.

Let a single household have the following tastes

$$\theta c + (1 - \theta) \ln n, \text{ with } 0 < \theta < 1,$$

where c and n denote the household's consumptions of market and nonmarket goods. Write the tastes for a married household as

$$\theta \left(\frac{c}{2}\right) + (1 - \theta) \ln \left(\frac{n}{2^{\kappa}}\right), \text{ again with } 0 < \kappa, \theta < 1.$$

Observe that the married household's consumption of the market good, c, is divided through by 2. This gives consumption per person. Now, suppose that a married household earns twice as much as a single household and consumes twice as much. Since household consumption is divided by 2 they would not be better off than a single household. In a similar vein, the consumption of nonmarket goods is divided through by 2^{κ}. As will be seen soon, this implies that the amount of nonmarket time spent producing married household utility will also be divided by 2. Thus, a married household will not be better off than a single household because it can devote twice as much time to household production. Now, the cost of household inputs is fixed at q. A married household, which has two earners, will be able to better afford this fixed cost than a single household. This is the sole source of economies of scale in the consumption of goods in the analysis. Note that market and nonmarket goods in a married household are *public* goods. They are enjoyed by both partners in exactly the same way. This is in contrast to the Becker marriage model, presented later, where a person's own consumption is a private good that is not enjoyed by her or his partner. In this case the household's resources must be split somehow between the partners.

4.1 Single Households

A single household solves the following maximization problem:

$$S(w, q, d) \equiv \max_{l \le 1 - q/w} \left\{ \underbrace{\theta[w(1 - l) - q]}_{c} + (1 - \theta)(1 - \kappa) \ln d + (1 - \theta)\kappa \ln l \right\}.$$

The objective function on the right is obtained by substituting the household production function into the single household's utility function. Here $S(w, q, d)$ represents the *indirect utility function* for a single household. It gives the *maximal* level of utility that the household can realize when it faces the wage rate, w, and has a stock of household inputs, d, which is purchased at the price q. This indirect utility function plays an important role in the subsequent analysis. As will be seen, it can be solved for explicitly by using the solutions for l and c that arise from the above maximization problem. The constraint on l in the above maximization problem ensures that consumption, c, is nonnegative. The right-hand side of the constraint, $1 - q/w$, is the time left over after the person has worked enough to earn the money necessary to pay for the household inputs, d. Both the corner and interior solutions to the above optimization problem play important roles in what follows. (Again, the Mathematical Appendix contains a discussion of interior and corner solutions.) When the constraint on l binds, the corner solution will apply. It will be shown that the corner solution is more likely to hold when the cost of household inputs, q, is high relative to wages, w. In this situation, the household must work a lot in the market to cover basic living costs, as reflected by q. Hence, the corner solution will be relevant when the economy is not very developed. The interior and corner solutions for housework, l, are now discussed in turn.

4.1.1 The interior solution, $l < 1 - q/w$
The first-order condition (when there is an interior solution) for a single household is given by

$$\underbrace{\theta w}_{\text{MC}} = \underbrace{(1-\theta)\kappa\frac{1}{l}}_{\text{MB}}. \tag{4.1}$$

This type of first-order condition should be familiar by now. The right-hand side gives the marginal benefit of working a bit more at home, MB. The marginal benefit has two components. An extra unit of home work, l, will raise the production of home goods by the marginal product of the person's time, $MP_l = \kappa d^{1-\kappa} l^{\kappa-1}$. An extra unit of home goods leads to a gain in utility of $MU_n = (1-\theta)/n = (1-\theta)/(d^{1-\kappa} l^{\kappa})$. Therefore, the marginal benefit of spending an extra unit of time at home is $MP_l \times MU_n = \kappa d^{1-\kappa} l^{\kappa-1} \times (1-\theta)/(d^{1-\kappa} l^{\kappa}) = (1-\theta)\kappa/l = \text{MB}$. The left-hand side is the marginal cost, MC. By working an extra unit at home, the person loses w in wages, which leads to a loss of θw in utility terms.

It is easy to see that the interior solution for the household's labor supply at home, l, is given by

$$l = \frac{(1-\theta)}{\theta}\frac{\kappa}{w} \quad \text{(interior solution, housework)},$$

which implies that the solution for market work, $1 - l$, is

$$1 - l = 1 - \frac{(1-\theta)}{\theta}\frac{\kappa}{w} \quad \text{(interior solution, market work)}.$$

Not surprisingly, the following properties hold when the interior solution applies for the time spent working at home, l:

(1) Time spent at home, l, is a decreasing function of the wage rate, w.

(2) Time spent at home is rising in the weight that the person places on home goods relative to market ones, $(1-\theta)/\theta$.

(3) It is also increasing in κ, which measures the value of labor in household production.

4.1.2 The corner solution, $l = 1 - q/w$
At the corner solution for housework, the first-order condition (4.1) appears as

$$\underbrace{\theta w}_{\text{MC}} < \underbrace{(1-\theta)\kappa\frac{1}{l} = \frac{(1-\theta)\kappa}{1-q/w}}_{\text{MB}}.$$

This states that the marginal benefit, MB, of devoting an extra unit of time to nonmarket work is greater than its marginal cost, MC. Therefore, the individual would like to devote more time to nonmarket work, but cannot do so because of the necessity to cover the fixed cost q.

For the interior solution discussed above to make sense, the consumption of market goods, c, must always be nonnegative. This requires that $c = w(1 - l) - q \geq 0$. The single household must always supply enough labor, $1 - l$, to the market to ensure that this is the case. Thus, $1 - l \geq q/w$. The term q/w measures the amount of time spent working that it would take to cover the fixed cost q. The corner solution applies when the interior solution for market work, $1 - l = 1 - (1 - \theta)\kappa/(\theta w)$, is smaller than q/w. This implies that at the interior solution the household cannot cover the fixed cost q. Therefore, the corner solution for market work, $1 - l$, is characterized by

$1 - l = q/w$, when $1 - (1 - \theta)\kappa/(\theta w) < q/w$ (corner solution, market work).

This corner solution can be equivalently expressed in terms of housework, l, as

$l = 1 - q/w$, when $1 - (1 - \theta)\kappa/(\theta w) < q/w$ (corner solution, housework).

Rearranging the first-order condition for the corner solution gives the criteria on the right-hand side of the above two expressions specifying when the corner solution holds, of course. The corner solution is more likely to transpire when wages, w, are low or the fixed cost of household formation, q, is high. So, the corner solution is most likely to apply at the early stages of economic development. When the corner solution holds, the single household will be short (or constrained) on the time it spends at home, l, in the sense that less time will be spent at home relative to the unconstrained solution; i.e., $l = 1 - q/w < [(1 - \theta)/\theta](\kappa/w)$.

The following properties hold when the corner solution applies for the time spent working at home, l:

(1) Time spent at home, l, is a decreasing function of the fixed cost, q. This is easy to explain. The higher the fixed cost is, the more time the individual must spend in the market to cover it.

(2) Time spent in the home is an increasing function of the wage rate, w, which is in contrast with the interior solution. This transpires because when wages are high, the person needs to spend less time working in the market to cover the fixed cost.

4.1.3 The complete solutions for housework, market work, and consumption
By combining the two cases together, the complete solution for housework, l, is

$$l = \begin{cases} 1 - q/w, & \text{if} \quad 1 - (1-\theta)\kappa/(\theta w) < q/w \\ & \text{(corner solution, housework);} \\ (1-\theta)\kappa/(\theta w), & \text{if} \quad 1 - (1-\theta)\kappa/(\theta w) \geq q/w \\ & \text{(interior solution, housework).} \end{cases}$$

Likewise, the solutions for market work, $1 - l$, and consumption, c, are given by

$$1 - l = \begin{cases} q/w, & \text{if} \quad 1 - (1-\theta)\kappa/(\theta w) < q/w \\ & \text{(corner solution, market work);} \\ 1 - (1-\theta)\kappa/(\theta w), & \text{if} \quad 1 - (1-\theta)\kappa/(\theta w) \geq q/w \\ & \text{(interior solution, market work),} \end{cases} \tag{4.2}$$

and

$$c = \begin{cases} 0, & \text{if} \quad 1 - (1-\theta)\kappa/(\theta w) < q/w \\ & \text{(corner solution, consumption);} \\ w - (1-\theta)\kappa/\theta - q, & \text{if} \quad 1 - (1-\theta)\kappa/(\theta w) \geq q/w \\ & \text{(interior solution, consumption).} \end{cases}$$

The solutions for c and l will now be used to solve for the single household's indirect utility function, $S(w, q, d)$.

4.1.4 The single household's indirect utility function, $S(w, q, d)$
The solution for the single household's indirect utility function, $S(w, q, d)$, can be obtained by plugging the optimal solutions for c and l into the objective function in the above maximization problem. There will be two cases, one for the interior solution and one for the corner solution. In particular,

$$S(w, q, d) = \begin{cases} (1-\theta)(1-\kappa)\ln d + (1-\theta)\kappa \ln[1 - q/w], \\ \text{if } 1 - (1-\theta)\kappa/(\theta w) < q/w \\ \text{(corner solution);} \\ \\ \theta w - (1-\theta)\kappa - \theta q + (1-\theta)(1-\kappa)\ln d \\ + (1-\theta)\kappa \ln[(1-\theta)\kappa/(\theta w)], \\ \text{if } 1 - (1-\theta)\kappa/(\theta w) \geq q/w \\ \text{(interior solution).} \end{cases} \tag{4.3}$$

Taking stock of the situation, one can see that the single household's maximization exhibits the following properties:

(1) A single household's utility, $S(w, q, d)$, will be higher when it is unconstrained as opposed to when it is constrained. To understand why, suppose that the household is unconstrained. It is feasible for this household to set its labor supply in the market, $1 - l$, to the constrained level q/w. If it does this, then this household would have the same level of utility as the constrained household. In general, though, this will not be the solution that maximizes its utility. The fact that the unconstrained household doesn't pick the same level of labor supply as the constrained one implies that the former's utility must be higher than the latter's. This is easy to see directly, as the next two points make clear.

(2) The unconstrained single household enjoys utility from the consumption of market goods in the amount $\theta w - (1 - \theta)\kappa - \theta q > 0$, whereas the constrained single household has zero utility.

(3) The unconstrained single household also devotes more time to household production, $(1 - \theta)\kappa/(\theta w)$, than does the constrained single household, $1 - q/w$.

4.2 Married Households

The problem facing a married household is the same, but for one key difference: the married household has two units of time. Hence, its maximization problem appears as

$$M(w, q, d) \equiv \max_{l \leq 2 - q/w} \left\{ \theta \underbrace{[w(2 - l) - q]/2}_{c} + (1 - \theta)(1 - \kappa)\ln d + (1 - \theta)\kappa \ln(l/2) \right\}.$$

The objective function on the right-hand side is obtained by substituting the household production function into the married household's utility function. Observe that the married household's time spent in household production, l, is divided by 2, similar to their consumption of market goods, $w(2 - l) - q$. So, while they may be able to spend twice as much time in household production relative to the single household, there will not be a benefit to marriage on this account. The first-order condition (when there is an interior solution) associated with this problem is

$$\frac{\theta w}{2} = (1 - \theta)\kappa\frac{1}{l}.$$

Observe that this first-order condition is the same as the (interior) one for single households, but for the fact that the left-hand side is divided by 2. This implies that the married household will devote twice as much labor to home production as the single household, when the interior solution holds for both households. For simplicity in the subsequent analysis, assume the following condition:

Condition 1 (Married Household Affordability) $1 - (1 - \theta)\kappa/(\theta w) \geq q/(2w)$.

This condition guarantees that the married household will never be at the corner solution or that the household's consumption will be nonnegative. To see this, substitute the interior solution for l associated with the above first-order condition into the condition for nonnegative consumption, $w(2 - l) - q \geq 0$, and then rearrange terms. The above condition is virtually identical to the single household's condition for an interior solution to hold, except that the right-hand side is now divided by 2. This makes it more likely to hold; it is easier for a married household to cover the fixed cost of establishing a household, q, because they have two units of labor as opposed to one for the single household.

4.2.1   The solutions for housework, market work, and consumption   The married household's solution for housework, l, market work, $2 - l$, and consumption, c, are given by

$l = 2(1 - \theta)\kappa/(\theta w)$ (housework),

$2 - l = 2 - 2(1 - \theta)\kappa/(\theta w)$ (market work),

and

$c = 2w - 2(1 - \theta)\kappa/\theta - q$ (consumption).

The solutions for c and l will be used to obtain an expression for the married household's indirect utility function, $M(w, q, d)$.

4.2.2   The married household's indirect utility function, $M(w, q, d)$ Substitute the above optimal solutions for c and l into the objective function associated with married household's maximization problem. It is then easy to see that the indirect utility function for a married couple is

$$M(w, q, d) = \theta w - (1 - \theta)\kappa - \theta q/2 \qquad (4.4)$$
$$+ (1 - \theta)(1 - \kappa)\ln d + (1 - \theta)\kappa \ln[\frac{(1 - \theta)\kappa}{\theta w}].$$

4.3   The Decision to Marry or Not

The decision to marry or not is governed by two considerations—namely, economic and noneconomic ones. The economic one arises from the economy of scale in household consumption. The noneconomic ones are love and companionship in a relationship. Each of these two factors is now visited in turn.

4.3.1   Money   By comparing (4.3) with (4.4), one can see that there is an economic gain from marriage. Specifically,

$$M(w,q,d) - S(w,q,d) = \begin{cases} \theta w - (1-\theta)\kappa - \theta q/2 \\ +(1-\theta)\kappa\{\ln[(1-\theta)\kappa/(\theta w)] - \ln[1-q/w]\} > 0, \\ \text{if } 1 - (1-\theta)\kappa/(\theta w) < q/w \\ \text{(corner solution for single);} \\ \\ \theta q/2 > 0, \\ \text{if } 1 - (1-\theta)\kappa/(\theta w) \geq q/w \\ \text{(interior solution for single).} \end{cases}$$

$$(4.5)$$

What explains the economic gain from marriage? There are two cases to consider.

(1) *The single household is unconstrained.* The married couple benefits because they can split the fixed cost of maintaining a household across two individuals. This leads to a utility gain of $\theta q/2$.

(2) *The single household is constrained.* Next, focus on the situation in which the single household is constrained. Recall that in this situation the household is consuming zero market goods; i.e., $c=0$. By marrying, a single person can increase the utility from their consumption of market goods from $\theta c = 0$ to $\theta c = \theta w - (1-\theta)\kappa - \theta q/2 \geq 0$. The individual will also raise their utility from the consumption of nonmarket goods by $(1-\theta)\kappa\{\ln[(1-\theta)\kappa/(\theta w)] - \ln[1-q/w]\} \geq 0$. This occurs because a married household spends more time in household production than a single one. The utility differential between married and single life, $M(w,q,d) - S(w,q,d)$, is bigger for the constrained household as opposed to the unconstrained one. On this, recall that $S(w,q,d)$ was smaller for the constrained household.

4.3.2 Love To incorporate love and companionship, suppose that upon meeting a couple draws a bliss variable, b. The variable b can be positive or negative, and measures their degree of compatibility. Suppose that $b \in [\underline{b}, \overline{b}]$ is distributed across potential couples according to a uniform distribution. (See the Mathematical Appendix for the definition of the uniform distribution.) Assume that $\underline{b} < 0 < \overline{b}$. If $\underline{b} > 0$, then everybody would marry, which is an undesirable counterfactual property.

4.3.3 Yes or No? Should a couple marry or not? A prospective couple use the following criteria to decide this:

MARRY,   if $M(w, q, d) + b \geq S(w, q, d)$;
REMAIN SINGLE, if $M(w, q, d) + b < S(w, q, d)$.

(When $M(w, q, d) + b = S(w, q, d)$ the couple is indifferent with regard to the choice of marrying or not. So, it doesn't really matter in which category this case is placed.) There will be a threshold value for bliss, b^*, at which the couple is indifferent about marrying or not. It is defined by $b^* = -[M(w, q, d) - S(w, q, d)] < 0$. Thus, using (4.5),

$$
b^* = \begin{cases}
-\theta w + (1 - \theta)\kappa + \theta q/2 \\
\quad -(1 - \theta)\kappa\{\ln[(1 - \theta)\kappa/(\theta w)] - \ln[1 - q/w]\} < 0, \\
\quad \text{if } 1 - (1 - \theta)\kappa/(\theta w) < q/w \\
\quad \text{(corner solution for single)}; \\
-\theta q/2 < 0, \\
\quad \text{if } 1 - (1 - \theta)\kappa/(\theta w) \geq q/w \\
\quad \text{(interior solution for single)}.
\end{cases}
\tag{4.6}
$$

So, the decision to marry can be recast as

MARRY,   if $b \geq b^*$;
REMAIN SINGLE, if $b < b^*$,

where b^* is given by (4.6). Observe that the threshold level of b is *negative*. This implies that a person may marry another person whom they don't really love because of the economic incentives for marriage. It does not imply that there will be a lack of love in most marriages. The value for b^* will be lower when a single household is constrained.

4.3.4 The fraction of the population that is married
The fraction of women (or men) who are married is given by

$$\Pr[b \geq b^*] = 1 - U(b^*) = 1 - \frac{b^* - \underline{b}}{\overline{b} - \underline{b}} = \frac{\overline{b} - b^*}{\overline{b} - \underline{b}}.$$

Even though the threshold value for a marriage, b^*, is negative, this does not imply that the mean value for marital bliss is negative. Recall that a person will marry whenever $b \geq b^*$. Some people—hopefully most people—will draw a positive value for b. The mean value of marital bliss will be

$$\int_{b^*}^{\overline{b}} \frac{b}{\overline{b} - b^*} \, db = \frac{\overline{b} + b^*}{2}.$$

The lowest that b^* can be is $\underline{b}$, because $b \in [\underline{b}, \overline{b}]$. Recall that the mean for b is $(\overline{b} + \underline{b})/2$. Therefore, if this mean is positive, then most people will be in a happy marriage because

$$\frac{\overline{b} + b^*}{2} \geq \frac{\overline{b} + \underline{b}}{2} > 0.$$

4.4 From Economics to Romance
Is the above framework useful for explaining the decline in marriage that occurred over the last one hundred years? The answer is yes. To see this, break up the economic development process into three underlying forces: a rise in wages, an advance in labor-saving household technology, and a decline in the price for home inputs.

(1) *A Rise in Wages.* Imagine an economy that is very poor. Wages will be low. The cost of buying the goods going into household production will be high. At low wages a single person will be at the corner solution, $1 - l = q/w$, where he or she works a lot in the market; see (4.2). The threshold value for love required for a marriage will be very low. Most independent people will be married. It should be fairly common to see couples together who don't really love each other. The lower the market wage is, the bigger the gain from marriage will be since it allows a person to get more consumption, and the lower the threshold value for marriage will be. Specifically, it can be shown that

$$\frac{db^*}{dw} = -\theta + \frac{(1 - \theta)\kappa}{w - q} > 0,$$

where the sign of the above expression follows from the fact that in the constrained region $1 - (1 - \theta)\kappa/(\theta w) < q/w$. Hence, as wages rise, people become pickier about whom they marry. Additionally, as wages grow, the single person will eventually reduce her or his labor effort in the market from q/w to $1 - (1 - \theta)\kappa/(\theta w)$. At this stage, the single household is unconstrained, and the differential in utility between married and single life is no longer a function of the wage rate, as (4.5) shows. Recall that figure 1.13 in chapter 1 shows the continuous rise in wages that has occurred in the United States over the last century.

(2) *Labor-Saving Technological Progress in the Household Sector.* Again, suppose that the single household is at a corner solution. Let there be labor-saving technological progress in the household sector in the sense that κ declines. This implies that less weight is being applied to labor in household production than to goods. Using (4.6), it is easy to calculate that for the constrained household

$$\frac{db^*}{d\kappa} = -(1 - \theta)\{\ln[(1 - \theta)\kappa/(\theta w)] - \ln[1 - q/w]\} < 0.$$

Thus, a *fall* in κ will lead to a rise in b^*. Therefore, the fraction of the population choosing single life will grow when there is labor-saving technological progress in the household sector. This will not be a consideration when single households are unconstrained, since the solution for b^* in (4.6) does not then involve κ. The diffusion of labor-saving household technologies in the United States is charted in figure 1.14, chapter 1.

(3) *A Decline in the Price of Household Inputs.* Suppose that the cost of the inputs required for forming a household declines or that q falls. Then, single life will gain value relative to married life. The threshold value for b will rise, as (4.6) indicates. Specifically, in the constrained region

$$\frac{db^*}{dq} = \frac{\theta}{2} - \frac{(1 - \theta)\kappa}{w - q} < 0,$$

which again follows from the fact that $\theta - (1 - \theta)\kappa/[w - q] < 0$, while in the unconstrained region

$$\frac{db^*}{dq} = -\frac{\theta}{2} < 0.$$

Hence, people will become pickier about their mate as the price of home inputs falls; i.e., b^* will rise as q *drops*. So the mean value for marital bliss will rise as q falls. People will marry more for love than for money than they did in the past. Figure 1.15 in chapter 1 displays the rapid drop in the prices for households appliances that occurred in the United States since World War II.

5   Divorce

Divorce is a recent phenomenon. In yesteryear, it was prohibited. Given its ubiquitous presence in society today, it is important to amend the above framework to allow for divorce. This will be done first in a setting where remarriage is prohibited. Then it will be done in one in which remarriage is allowed. As will be seen, the possibility of remarriage raises the value of a divorce.

5.1   England, 1660–1857

Divorce was not always as easy as it is today.[4] Marriage in England could be like quicksand: easy to walk into and impossible to get out of. A verbal contract between a man and a woman that was witnessed by at least two people was all the legal cement that was needed to glue a marriage together. Marital matters were largely adjudicated by ecclesiastical courts. The Church of England interpreted the words of Christ as meaning that a marriage was indissoluble. "Till death do us part" had real meaning. An official system of legalized divorce was not adopted until the Divorce Act of 1857.

So how did one get out of a marriage? Among the underclass many marriages were held in secrecy, presumably to facilitate subsequent breakups. One-third of London plebeian marriages took place clandestinely in the Fleet Prison.[5] Clandestine marriages were forbidden in 1754. Another solution was to just to walk away. Desertion was widespread, usually by men. Men often ran off and set up another household or joined the military. Bigamy was common, even though in theory it could be punished by death; in practice, the offender's hand would be burned.

For the propertied the situation was more complicated since the option of disappearing was unattractive. A private separation agreement could be

4. The English marital system is detailed in Stone (1993).
5. See Kent (1990).

drawn up, if there was mutual agreement between the parties. This agreement specified three things: the amount of alimony that the wife was due, the wife's financial independence, and the custody of the children. Remarriage was still not legally possible. Generally a wife was entitled to about one-third of her husband's income. Making a delinquent husband pay could be extremely difficult, just as it is today. Under common law a husband was responsible for the family's financial affairs. So, without an agreement giving the wife financial independence, the husband would still have rights to his wife's future income streams or be liable for her future debts. Last, the husband had absolute and inalienable rights to the children, which meant that any agreement concerning custody of the children was not legally enforceable. In a society based on primogeniture, wealthy husbands would be reluctant to give up their sons. And custody rights were often used as bargaining chips by husbands to reduce alimony settlements. Another alternative was to litigate a settlement through the courts. This was an option open mainly to the wealthy. The sole grounds for judicial separations were life-threatening cruelty or adultery. The husband still had rights to and liabilities for the family's financial affairs.

5.2   The Decision to Divorce
The decision to divorce or not is analogous to the decision to get married. To see this, extend the above model to two periods. Let a household now have tastes given by

$$\theta c + (1 - \theta) \ln n + \beta[\theta c' + (1 - \theta) \ln n'], \text{ with } 0 < \beta, \theta < 1,$$

where the prime attached to a variable denotes its value in the second period. The variable β represents the discount factor. People value future utility flows less than current ones; i.e., $0 < \beta < 1$. Events are timed as follows. At the beginning of the first period every person meets someone of the opposite sex. Each prospective couple draws a bliss quality of value b from the uniform distribution $U(b)$ on $[\underline{b}, \overline{b}]$. They then decide to marry or not. They do this on the basis of romantic interest, b, and economic factors such as the value of wages, w, and the price of durables, q. (Recall from the earlier analysis that the decision to marry did not depend on the value of d. This is also true in what follows.) If they marry, they then enter the second period as a married couple and draw a new value for the bliss variable, b', again from the distribution $U(b')$. At this time the value of wages and the price of durables will be w' and q'. The couple will decide whether or not to remain married on the basis of b', w', and q'. Divorce is assumed to be costless. If the couple does not marry in the first period, then each person will meet someone else in the second period. The new couple will draw a bliss quality, b', and decide whether to marry for the last period. Suppose that divorcees cannot remarry. (The possibility of remarriage is introduced in the next section.)

Now, for concreteness, define the following indirect utility functions specifying the values of single life in the first and second periods:

$$S(w, q, d) \equiv \max_{l} \{\theta[w(1-l) - q] + (1-\theta)(1-\kappa)\ln d + (1-\theta)\kappa \ln l\},$$

and

$$S(w', q', d') \equiv \max_{l'} \{\theta[w'(1-l') - q'] + (1-\theta)(1-\kappa)\ln d' + (1-\theta)\kappa \ln l'\}.$$

(Note that the expressions for S assumes that the interior solution for a single person's time at home holds.) Likewise, the indirect utility functions for married life in the first and second periods are

$$M(w, q, d) \equiv \max_{l} \{\theta[w(2-l) - q]/2 + (1-\theta)(1-\kappa)\ln d + (1-\theta)\kappa \ln(l/2)\},$$

and

$$M(w', q', d') \equiv \max_{l'} \{\theta[w'(2-l') - q']/2 + (1-\theta)(1-\kappa)\ln d' + (1-\theta)\kappa \ln(l'/2)\}.$$

When will a divorce happen? Imagine that a married couple is in the second period of their lives. The couple will terminate their marriage if

$$M(w', q', d') + b' < S(w', q', d').$$

There will be a threshold value for b', given by $b'^* = -[M(w', q', d') - S(w', q', d')]$, such that the couple will

DIVORCE, if $b' < b'^*$;

REMAIN MARRIED, if $b' \geq b'^*$.

There could be several reasons for divorce. First, perhaps "the thrill is gone." To see this, let $b'^* = b^*$, so that the thresholds for the two periods are same, which would occur when w, q, and d are constant over the time. Now, suppose that the couple draws a b' below b^*. They will now divorce. Second, the economic value of a marriage, relative to single life, may have declined, which would result in a rise in b'^*. This could be due to a rise in wages ($w' > w$), labor-saving technological progress in the household sector

$(\kappa' < \kappa)$, and a fall in the price of household inputs $(q' < q)$. Now, the rate of divorce (out of the stock of people who got married in the first period) is given by

$$\Pr[b' < b'^*] = U(b^*) = \frac{b'^* - \underline{b}}{\overline{b} - \underline{b}}.$$

Therefore, the long-run rise in the rate of divorce is driven by growth in wages, labor-saving advances in household technology, and drops in the price of household inputs.

It is easy to put a cost of divorce into the analysis. Let a divorce cost λ, which is split equally across both parties. Then, the above condition for a divorce becomes

$$M(w', q', d') + b' < S(w', q', d') - \theta \frac{\lambda}{2}.$$

This happens because the need to pay for a divorce will reduce the utility from market consumption for each person by $\theta \lambda / 2$. It is easy to see that introducing a divorce cost will lower the threshold value of bliss to stay married, b'^*, since now $b'^* = -[M(w', q', d') + \theta\lambda/2 - S(w', q', d')]$. Hence, there will be fewer divorces. California introduced no-fault divorce in 1969, which reduced the cost of a divorce. Other states quickly followed.

5.3   Remarriage

How does the possibility of remarriage affect the above analysis? Specifically, suppose that if a married couple gets divorced at the beginning of the second period, then they too can go into the singles market and meet a potential mate. This will influence the decision to divorce. The fact that a single divorcee can potentially find a new mate increases the value of divorce.

Imagine that two divorcees meet on the marriage market. Let $\widehat{b}'$ represent the value of bliss this couple will draw. This is drawn from the uniform distribution specified earlier on $[\underline{b}, \overline{b}]$. They will marry if

$$M(w', q', d') + \widehat{b}' \geq S(w', q', d').$$

The threshold value of bliss, $\widehat{b}^{*\prime}$, for the divorcees is given by

$$\widehat{b}^{*\prime} = S(w', q', d') - M(w', q', d') < 0.$$

Therefore, the couple will

REMARRY, if $\widehat{b}' \geq \widehat{b}^{*\prime}$;

REMAIN DIVORCEES, if $\widehat{b}' < \widehat{b}^{*\prime}$.

What are the odds that $\widehat{b}' \geq \widehat{b}^{*\prime}$? This gives the odds that a divorcee will remarry, which are given by

$$\rho \equiv \Pr(\text{remarriage}|\text{divorce}) = \Pr(\widehat{b}' \geq \widehat{b}^{*\prime}) = 1 - \Pr(\widehat{b}' \leq \widehat{b}^{*\prime})$$

$$= 1 - \frac{\widehat{b}^{*\prime} - \underline{b}}{\overline{b} - \underline{b}} = \frac{\overline{b} - \widehat{b}^{*\prime}}{\overline{b} - \underline{b}}.$$

The expected value of $\widehat{b}'$ for a remarriage is

$$\frac{\overline{b} + \widehat{b}^{*\prime}}{2}.$$

Imagine the decision facing a married couple at the beginning of the second period. They know that if they divorce, then they will remarry with probability ρ and remain single with probability $1 - \rho$. If they remarry, they can expect a marital bliss of $(\overline{b} + \widehat{b}^{*\prime})/2$. So, the couple will terminate their marriage in the second period if

$$M(w', q', d') + b' < \rho\left[M(w', q', d') + \frac{\overline{b} + \widehat{b}^{*\prime}}{2}\right] + (1 - \rho)S(w', q', d').$$

The left-hand side of the above equation gives the utility from remaining married. The right-hand side represents the *expected* utility from a divorce. This has two components. The first is the expected utility from a remarriage, which is $M(w', q', d') + (\overline{b} + \widehat{b}^{*\prime})/2$. A remarriage occurs with probability ρ. The second is the utility from single life, $S(w', q', d')$, which happens with odds $1 - \rho$.

The above equation can be rewritten as

$$\rho\left(\frac{\overline{b} + \widehat{b}^{*\prime}}{2} - b'\right) > (1 - \rho)[M(w', q', d') + b' - S(w', q', d')].$$

This equation has a nice interpretation. Focus on the right-hand side first. If $M(w', q', d') + b' - S(w', q', d') < 0$, then the couple will divorce for sure. Here, the bliss in the current marriage is so low that a person would prefer to be single than stay in the current marriage. Note that in this situation $b' < 0$, because $M(w', q', d') > S(w', q', d')$. Since $b' < 0$, the left-hand side is positive. Now take the case where $M(w', q', d') + b' - S(w', q', d') > 0$, so that there is some value from the current marriage. A person may still divorce, if the expected gain from finding someone better, $\rho[(\bar{b}+\widehat{b}^{*\prime})/2 - b']$, exceeds the expected loss from terminating the current marriage and remaining single, $(1-\rho)[M(w',q',d') + b' - S(w',q',d')]$.

The threshold value of bliss for a divorce (or for b'), denoted by $b^{*\prime}$, is given by

$$b^{*\prime} = (1-\rho)[S(w',q',d') - M(w',q',d')] + \rho\left(\frac{\bar{b}+\widehat{b}^{*\prime}}{2}\right)$$

$$= (1-\rho)\widehat{b}^{*\prime} + \rho\left(\frac{\bar{b}+\widehat{b}^{*\prime}}{2}\right) > \widehat{b}^{*\prime}.$$

Interestingly, $b^{*\prime} > \widehat{b}^{*\prime}$. (Note that $(1-\rho)\widehat{b}^{*\prime} + \rho(\bar{b}+\widehat{b}^{*\prime})/2 > \widehat{b}^{*\prime}$, because $\rho(\bar{b}-\widehat{b}^{*\prime})/2 > 0$.) Therefore, the threshold value of bliss for a divorce exceeds that for a remarriage. The person considering a divorce realizes that there will still be one more draw on the marriage market.

6 Assortative Mating

To analyze assortative mating, the basic model of marriage developed in section 4 will be used, with just a few modifications. Each sex has one unit of time. To keep things simple, suppose that a man spends all of his time working in the market while a married woman divides her time between market work and household production. To have assortative mating men and women need to differ along some dimensions. Assume that there are two types of men, those with low productivity working on the labor market and those with high productivity. Denote a man's productivity level by μ_i for $i=1,2$, with $\mu_2 > \mu_1$. A man with productivity μ_i earns $w\mu_i$ on the market. Let there be π_i men who have a productivity level of μ_i. Similarly, suppose that women differ in their labor market productivity as well. Represent a woman's productivity in the market by ϕ_j for $j=1,2$, with $\phi_2 > \phi_1$. A woman with productivity ϕ_j receives compensation in the amount $w\phi_j(1-l)$, if she works $1-l$ hours in the market. Additionally, let women also differ in their productivity at home, η_h, for $h=1,2$, with $\eta_2 > \eta_1$. A woman with home productivity η_h supplies $\eta_h l$ units of effective household labor, if she works l hours at home. There are χ_{jh} women with a market productivity of ϕ_j and a home productivity of η_h. Thus, there are two types of men and four types of women. Normalize the total number of people of each sex to one; i.e., let $\pi_1 + \pi_2 = 1$ and $\chi_{11} + \cdots + \chi_{22} = 1$. For simplicity, assume that $\chi_{22} < \pi_2$ and $\chi_{11} < \pi_1$.

Who will marry whom in this economy? For the moment, assume that all matches are based solely on economic considerations. Clearly, not all men can marry a woman of the ideal type (ϕ_2, η_2). There are not enough of these women for even the best type of men, since $\chi_{22} < \pi_2$. Likewise, not all women can obtain a husband of the best type, because $\mu_2 < 1$. How should matching be done in this economy?

6.1 The Gale-Shapley Matching Algorithm

Focus on a marriage between a type-μ_i man and a type-(ϕ_j, η_h) woman. Their decision problem will appear as

$$M(\mu_i, \phi_j, \eta_h, w, q, d) \equiv \max_l \{\theta[w\mu_i + w\phi_j(1 - l) - q]/2 + (1 - \theta)(1 - \kappa)\ln d$$

$$+ (1 - \theta)\kappa \ln \eta_h + (1 - \theta)\kappa \ln(l/2)\}. \qquad (6.1)$$

This married household's problem is analogous to the one developed in section 4, so it will not be explained. The first-order condition for the married woman's effort at home reads

$$\frac{\theta w\phi_j}{2} = (1 - \theta)\kappa\frac{1}{l}. \qquad (6.2)$$

Assume that an interior solution always prevails. A woman will work more in the market the higher her productivity, ϕ_j, is there. Plugging this solution for l into the objective function then gives a solution for the indirect utility function, $M(\mu_i, \phi_j, \eta_h, w, q, d)$. Specifically, one obtains:

$$M(\mu_i, \phi_j, \eta_h, w, q, d) = \theta(w\mu_i + w\phi_j - q)/2 - (1 - \theta)\kappa \qquad (6.3)$$

$$+ (1 - \theta)(1 - \kappa)\ln d + (1 - \theta)\kappa \ln \eta_h$$

$$+ (1 - \theta)\kappa \ln\left[\frac{(1 - \theta)\kappa}{\theta w\phi_j}\right].$$

The indirect utility function provides the (maximized) value for a type-(μ_i, ϕ_j, η_h) marriage. By applying the envelope theorem (discussed in chapter 1) to the above maximization problem (6.1), or by direct calculation using (6.2) and (6.3), one can see that:

$$\frac{dM(\mu_i, \phi_j, \eta_h, w, q, d)}{d\mu_i} = \theta w/2 > 0, \qquad (6.4)$$

$$\frac{dM(\mu_i, \phi_j, \eta_h, w, q, d)}{d\phi_j} = \theta w(1-l)/2 > 0,$$

$$\frac{dM(\mu_i, \phi_j, \eta_h, w, q, d)}{d\eta_h} = (1-\theta)\kappa \frac{1}{\eta_h} > 0.$$

Therefore, a women will desire a husband with the high market productivity, μ_2, other things being equal. Likewise, a man will desire a wife who is productive in the labor force. That is, he prefers one who has ϕ_2 over ϕ_1. He would also like one who is productive at home or who has η_2. It is unclear how he would trade off a woman's productivity in the market for productivity at home. This will depend upon the state of development in the economy.

To characterize the implied matching process, simply make a list of the marital utilities from all possible pairings, starting from the top and going down to the bottom. The best women will be matched with the best men. Recall that there are more of these men than women; i.e., $\pi_2 > \chi_{22}$. Therefore, some of the high-type men will have to match with the next best women on the list. The matching process continues down this list in this fashion. At each stage the remaining best men are matched with the remaining best women. If there is an excess supply of one of the sexes, the overflow of this sex must find a match on the next line(s) of the list.

Now, suppose that the k-th position on the list is represented by a match of type (μ_i, ϕ_j, η_h). Some type-μ_i men may have already been allocated to women who are higher on the list—i.e., to women who have a better combination of ϕ_j and η_h. Let $R_m^k(\mu_i)$ be the amount of remaining type-μ_i men who can be allocated at the k-th position on the list. Similarly, let $R_f^k(\phi_j, \eta_h)$ be the number of available type-(ϕ_j, η_h) women. The number of matches is given by $\min\{R_m^k(\mu_i), R_f^k(\phi_j, \eta_h)\}$. This can be zero, since there may be no men of type-i left and/or no women of type (ϕ_j, η_h). Recall that the number of people of each sex is one. Thus, the odds of a type-(μ_i, ϕ_j, η_h) match are $\Pr(\mu_i, \phi_j, \eta_h) = \min\{R_m^k(\mu_i), R_f^k(\phi_j, \eta_h)\}$. Any type-$\mu_i$ men who are not assigned a mate at position k will be available for position $k+1$, and likewise for type-(ϕ_j, η_h) women. Thus, the number of type-μ_i men who will be available for the next position, $k+1$, will be given by $R_m^{k+1}(\mu_i) = R_m^k(\mu_i) - \min\{R_m^k(\mu_i), R_f^k(\phi_j, \eta_h)\}$, while the number of type-(ϕ_j, η_h) women is $R_f^{k+1}(\phi_j, \eta_h) = R_f^k(\phi_j, \eta_h) -$

Table 3.2
Gale-Shapley matching algorithm.

Ranking	Utility	Odds
1	$M(\mu_2, \phi_2, \eta_2, w, q, d)$	$\Pr(\mu_2, \phi_2, \eta_2) = \chi_{22}$
$\vdots$	$\vdots$	$\vdots$
k	$M(\mu_i, \phi_j, \eta_h, w, q, d)$	$\Pr(\mu_i, \phi_j, \eta_h) = \min\{R_m^k(\mu_i), R_f^k(\phi_j, \eta_h)\}$
$\vdots$	$\vdots$	$\vdots$
8	$M(\mu_1, \phi_1, \eta_1, w, q, d)$	$\Pr(\mu_1, \phi_1, \eta_1) = \min\{R_m^8(\mu_1), R_f^8(\phi_1, \eta_1)\} = \chi_{11},$

$\min\{R_m^k(\mu_i), R_f^k(\phi_j, \eta_h)\}$. The above iterative procedure begins with the starting conditions $R_f^1(\phi_j, \eta_h) = \chi_{jh}$ and $R_m^1(\mu_i) = \mu_i$, for all $h, i,$ and j. The matching process is then summarized by table 3.2.

The odds of some of the matches happening in table 3.2 are zero. For example, consider a marriage of type (μ_2, ϕ_1, η_1). There will be no type-μ_2 men left by the time the algorithm reaches a woman of type-(ϕ_1, η_1).

Last, a person at a particular rank in the table will be unable to entice a person of the opposite sex at a higher rank to enter into a marriage with them. The value from the marriage that the person at a lower rank brings is less than the value of the marriage that the person of a higher rank currently has. The above matching scheme is stable in the sense that any reshuffling of marriages must make someone worse off.

Example 1 (The Gale-Shapley Matching Algorithm) *Assume that a marriage with a type-μ_2 man is always better than a marriage with a type-μ_1 man. Now, for a given type of man, presume that a marriage with a type-(ϕ_2, η_1) woman delivers a higher level of marital bliss than with a type-(ϕ_1, η_2) woman. Thus, women are ranked as follows: $(\phi_2, \eta_2) > (\phi_2, \eta_1) > (\phi_1, \eta_2) > (\phi_1, \eta_1)$. For simplicity's sake, assume that $\chi_{21} + \chi_{22} = \pi_2$ and $\chi_{11} + \chi_{12} = \pi_1$. This implies that there is an equal number of men and women who have the high (low) productivity in the market. The matches that arise are shown in table 3.3.*

The highest valued marriage occurs when a type-μ_2 man matches with a type-(ϕ_2, η_2) female. Not all type-μ_2 men can be matched with type-(ϕ_2, η_2) women, though, because these women are in short supply. Hence, only χ_{22} such matches can occur. The remaining type-μ_2 men must be matched with the next-best women, here type (ϕ_2, η_1). At this stage, there are no type-μ_2 men left. There are

Table 3.3
Matching algorithm, example.

Ranking	Utility	Odds
1	$M(\mu_2, \phi_2, \eta_2, w, q, d)$	$\Pr(\mu_2, \phi_2, \eta_2) = \chi_{22}$
2	$M(\mu_2, \phi_2, \eta_1, w, q, d)$	$\Pr(\mu_2, \phi_2, \eta_1) = \pi_2 - \chi_{22} = \chi_{21}$
3	$M(\mu_2, \phi_1, \eta_2, w, q, d)$	$\Pr(\mu_2, \phi_1, \eta_2) = 0$
4	$M(\mu_2, \phi_1, \eta_1, w, q, d)$	$\Pr(\mu_2, \phi_1, \eta_1) = 0$
5	$M(\mu_1, \phi_2, \eta_2, w, q, d)$	$\Pr(\mu_1, \phi_2, \eta_2) = 0$
6	$M(\mu_1, \phi_2, \eta_1, w, q, d)$	$\Pr(\mu_1, \phi_2, \eta_1) = 0$
7	$M(\mu_1, \phi_1, \eta_2, w, q, d)$	$\Pr(\mu_1, \phi_1, \eta_2) = \chi_{12}$
8	$M(\mu_1, \phi_1, \eta_1, w, q, d)$	$\Pr(\mu_1, \phi_1, \eta_1) = \pi_1 - \chi_{12} = \chi_{11}$

also no more type-(ϕ_2, η_2) and type-(ϕ_2, η_1) women left either. All that remains are type-μ_1 men who must be allocated across type-(ϕ_1, η_2) and type-(ϕ_1, η_1) women. So, the algorithm must jump down to position 7 on the list. Since type-(ϕ_1, η_2) women are in short supply, only χ_{12} men can be matched with them. The rest, $\pi_1 - \chi_{12} = \chi_{11}$, will be assigned a type-(ϕ_1, η_1) wife.

6.2 Putting Some Randomness Back in Matching

A drawback of the above mating scheme is that the matches are perfectly assortative or predictable. In the real world they are far from this, as the contingency tables in section 3.4 illustrate. Other considerations enter into a marriage decision, such as love and companionship. Any degree of assortative mating in the economy can be obtained by assuming that a fraction β of each type mates in the above fashion while the remaining fraction, $1 - \beta$, matches randomly. Think of this as being two separate subpopulations in the economy, each with the same type distributions. With random matching, the probability that a type-μ_i man will marry a type-(ϕ_j, η_h) woman, or $\Pr^r(\mu_i, \phi_j, \eta_h)$, is given by $\Pr^r(\mu_i, \phi_j, \eta_h) = \pi_i \chi_{jh}$. Thus, by summing over the two separate subpopulations, the fraction of matches in the economy at large that are of type-(μ_i, ϕ_j, η_h) is

$$\beta \Pr(\mu_i, \phi_j, \eta_h) + (1 - \beta) \Pr^r(\mu_i, \phi_j, \eta_h).$$

Therefore, as $\beta \to 1$, the matchings will appear to be perfectly assortative, while as $\beta \to 0$, they will appear random. One could think about the random matches as being generated by a random bliss variable, b. Let $b \in \{0, \infty\}$ with $\Pr[b = 0] = \beta$ and $\Pr[b = \infty] = 1 - \beta$. One interpretation of this is that with probability $1 - \beta$ a person will find a mate drawn at random from the population whom they will marry no matter what. Love dictates this decision. With probability β they don't meet such a mate. Then, they match with the remaining population according to the Gale-Shapley algorithm. Here economic considerations guide the match.

6.3 The Rise in Positive Assortative Mating

What can explain the rise in positive assortative mating? Once again, entertain the hypothesis that labor-saving technological progress in the home and a rise in real wages are the drivers of this development. If these drivers reshuffle the order of the entries in table 3.2, then a change in the pattern of assortative mating will occur.

(1) *Yesteryear.* Suppose that women do little work in the market. A high value for η implies that a woman will have a high ranking in table 3.2. Specifically, assume that

$$M(\mu_i, \phi_1, \eta_2, w, q, d) > M(\mu_i, \phi_2, \eta_1, w, q, d), \text{ for all } \mu_i.$$

Here all men prefer a woman who is good at home production and not good at market work to a woman who is not good at home production and good at market work (note that the above condition holds for all μ_i). For the above condition to occur, it can be seen from equation (6.3) that the following expression must hold:

$$\underbrace{[w\theta(\phi_1 - \phi_2)/2 - (1-\theta)\kappa(\ln\phi_1 - \ln\phi_2)]}_{<0} + \underbrace{[(1-\theta)\kappa(\ln\eta_2 - \ln\eta_1)]}_{>0} > 0.$$

The first term in the above expression must be negative. To see this, recall that the indirect utility function is increasing in ϕ_j by (6.4). By inspecting (6.3), this implies that $w\theta\phi_1/2 - (1-\theta)\kappa\ln\phi_1 < w\theta\phi_2/2 - (1-\theta)\kappa\ln\phi_2$. The second term in the expression is obviously positive. It's clear that by making $\ln\eta_2 - \ln\eta_1$ large enough, the second term can be made to dominate. Therefore, there must exist situations where a man prefers a woman who is good at home production relative to one who is good at market work, if he must choose between these two alternatives. The next two cases address situations where the above expression may switch sign and become negative as a result of economic development.

(2) *Technological Progress in the Home.* Recall that κ is the exponent on labor in the household production function, while $1 - \kappa$ is the one on

goods. As κ falls and $1 - \kappa$ rises, labor becomes less important at home. Clearly, if κ falls due to labor-saving technologies in the home, then the expression could become negative. This would happen when $\kappa \simeq 0$, for example. Then, a man would prefer a wife who is not good at home production and good at market work to the wife who is good at home production and not good at market work. The ranking of $M(\mu_i, \phi_1, \eta_2, w, q, d)$ vis-à-vis $M(\mu_i, \phi_2, \eta_1, w, q, d)$ on the list would be reversed.

(3) *A Rise in Wages.* A large enough increase in wages would also cause the above expression to become negative. This makes sense because the value of a woman's work on the market is measured by w. When w is high, the value of marrying a woman who is productive on the market will also be high.

6.3.1 Measuring the strength of assortative mating empirically   How does this relate to the finding discussed in section 3.4 that assortative mating has risen over time? With the Gale-Shapley algorithm, mating is always perfectly assortative by construction. Since μ_i and ϕ_j reflect men's and women's productivities in the market, they will be associated with education or income levels. A person's education and/or income levels are observed by econometricians and statisticians. A woman's productivity at home, η_h, is not. Hence, the statistician is restricted to measuring assortative mating along only those dimensions that can be observed. This may lead the statistician to conclude that mating is not perfectly assortative. For example, think about the case just discussed, where $M(\mu_i, \phi_1, \eta_2, w, q, d) > M(\mu_i, \phi_2, \eta_1, w, q, d)$ for all i; i.e., here it might seem like assortative mating is violated because a type-μ_2 man prefers a type-ϕ_1 woman to a type-ϕ_2 one, because the statistician doesn't observe the hidden η dimension. (Additionally, another hidden dimension is the bliss or love that was discussed earlier.) Thus, the reshuffling in the order of the entries in table 3.2 just analyzed will look like a rise in assortative mating, given that the statistician observes only μ_i and ϕ_j.

7   Growing Up with a Single Mother

At the present time never-married adults need not find mates and marry for economic reasons, as in the past. They can wait until they find partners whom they truly love. Likewise, people no longer have to remain in an unhappy marriage. They can divorce and live alone until they potentially find someone else. There is one downside, though. Children from single-parent families are less likely to be successful than children living with two parents. This is the subject now.

7.1 Background

In a classic book, McLanahan and Sandefur (1994) documented that children growing up in single-parent families are worse off, *statistically speaking*, than those from two-parent ones. The study is based on an analysis of the large amounts of information contained in four socioeconomic data bases: the Panel Study of Income Dynamics (PSID), the National Longitudinal Survey of Young Men and Women (NLSY), the High School and Beyond Study (HSB), and the National Survey of Families and Households (NSFH). The PSID oversamples poor families, while the NLSY does the same for blacks and Hispanics.[6] This makes them ideal for studying the effects of poverty. Table 3.4 updates some of their statistics. As can be seen, children living in single-parents households are more likely than children from two-parent families to drop out of high school (15 percent vs. 9 percent), to be idle (26 vs. 12 percent), and to experience teen births (33 vs. 21 percent); they are less likely to go to college (15 vs. 24 percent). What economic factors might be important in accounting for these differences? Poverty is the first thing that comes to mind. Poverty is both a cause and symptom of single parenthood. It is a symptom because single parents have less wherewithal, in terms of money and time, to invest in their children relative to two-parent families. It is also cause in that children from single-parents families have less investment in them. Hence, they will be less successful in life, generally, and more likely to become single parents themselves. Poverty begets poverty, so to speak.

6. Oversampling is a statistical technique to make the survey results more accurate for smaller subgroups of the population. This is done by surveying proportionately more members of this group than the general population. The statistical sampling error for small groups is reduced as a result. (As an extreme example, imagine calling 10,000 people for a survey on wealth and savings. Rich people might save very differently than poor ones. How many billionaires would you get, and if by chance you got one, would the results be representative of the billionaire population?) The results for the oversampled subgroup are weighted in the survey according to their correct representation in the population. So, oversampling does not imply that these subgroups are overrepresented in the survey.

Table 3.4
U.S. child well-being by family structure.

	Single parent, %	Two parent, %
Relative income	51.4	·100
<High school	14.6	9.2
College	15.0	23.6
Teen births	32.8	21.3
Idle, at age 19–21	25.9	12.1

Source: Greenwood, Guner, and Vandenbroucke 2017.

Table 3.5
Total spending on children as a fraction of household income in the United States.

	(Households < $61,530)	
Child's age	Two parent (mean = $39,360)	Single parent (mean = $27,290)
0–2	0.24	0.30
3–5	0.24	0.33
6–8	0.23	0.32
9–11	0.25	0.35
12–14	0.26	0.36
15–17	0.26	0.35

Source: Greenwood, Guner, and Vandenbroucke 2017.

Children living with a single mother are much more prone to live in poverty than those living within a married household, as the right-hand-side panel of figure 3.6 shows. A much greater percentage of female-headed families receives welfare than married ones (see the left-hand-side panel of figure 3.6). Not surprisingly, for families in the bottom part of the income distribution (< $61,530) the cost of raising children represents a large fraction of their income (see table 3.5). This is especially true for single-parent families. Additionally, families from higher socioeconomic groups spend more time with their children. College-educated mothers spend more time with their kids than less-than-college educated ones (table 3.6). Single mothers spend less time with their offspring than do married ones. All in all, single-parent families have fewer resources, both in terms of time and money, to invest in their kids.

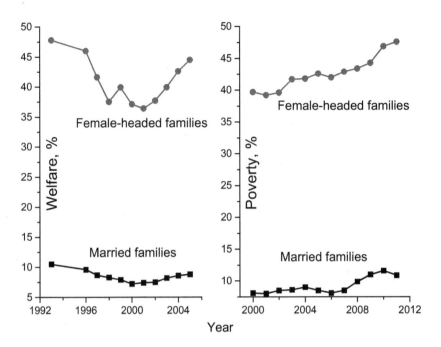

Figure 3.6
U.S. households on welfare (left panel) and the percentage of children living in poverty (right panel).
Source: Greenwood, Guner, and Vandenbroucke 2017.

7.2 The Model

Imagine a world where there are two types of men and women, those with a high level of productivity, $\phi_2/2$, and those with a low level, $\phi_1/2$, so that $\phi_1 < \phi_2$. Normalize the size of the populations of men to be one and likewise do the same for women. Men and women meet on a marriage market. They sort perfectly according to their productivity so that both husband and wife are from the same socioeconomic class. The couple draws a marital bliss shock, $\tilde{b}$, contained in $\{-b, b\}$. The realized value of the shock will take the low value, $-b < 0$, with probability β. The couple decide whether or not to marry based upon the realized value of this shock. Regardless of whether they marry or not, each woman has two children, a boy and a girl. Children always live with their mothers. An adult's productivity level is influenced by the amount of human capital investment that he or she received when young. In particular, if a household invests η in their children's education, then *both* children will draw

Table 3.6
Child care time by mothers in the United States.

Year	Hours per week			
	<Col.	Col.	Unmar.	Mar.
	Total			
1965	8.56	10.81	6.02	10.31
1975	7.73	7.38	6.80	7.94
1985	7.65	8.82	5.76	8.77
2003	12.51	16.39	11.14	14.41
2013	13.00	15.63	11.51	14.5
	Per child			
1965	3.38	5.09	2.82	4.71
1975	4.02	3.63	3.98	3.90
1985	4.67	5.38	3.72	5.27
2003	7.24	9.63	7.38	8.04
2013	7.76	9.39	7.42	8.47

Source: Greenwood, Guner, and Vandenbroucke 2017.

a high level of productivity, $\phi_2'/2$, with probability π. It is assumed that $\eta < \phi_1/2$, so that it is feasible for all households to educate their children. Here a prime is attached to ϕ to denote the productivity of the children as opposed to the parents. If they don't invest in their children's education, then both children will draw the high level with probability $\lambda < \pi$.

Let a married household have preferences of the form

$$u = \ln c + 2\phi' + \tilde{b},$$

where c is consumption, $\phi' \in \{\phi_1/2, \phi_2/2\}$ represents the productivity level of the children, and $\tilde{b} \in \{-b, b\}$ denotes the level of marital bliss. Parents care about the success of their children, as measured by their offsprings' productivity. Utility for a single mother is given by

$$u = \ln c + 2\phi'.$$

She does not realize any bliss from marriage. The preferences for a single man have the form

$$u = \ln c.$$

Since the children do not live with their fathers, the model assumes that fathers do not enjoy a benefit from their children and consequently will not invest in them.

7.3   Investment in Human Capital

Start with a married household. Should they invest in their children? Let $h = 0$ indicate, if they don't, and $h = 1$, if they do. If they don't invest in their children, then a type-i household's consumption will be ϕ_i, for $i = 1, 2$. The *expected* utility from children is then $(1 - \lambda)\phi_1' + \lambda\phi_2'$. Alternatively, if they do invest, then the household's consumption will be $\phi_i - \eta < \phi_i$. The expected utility from their children will be $(1 - \pi)\phi_1' + \pi\phi_2' > (1 - \lambda)\phi_1' + \lambda\phi_2'$. Thus, investing in children reduces the household's consumption, but increases the expected utility that the parents will realize from their kids. The decision to invest for a type-i married household (for $i = 1, 2$) is summarized by

$$h = \begin{cases} 0, & \text{if } \ln(\phi_i - \eta) + (1 - \pi)\phi_1' + \pi\phi_2' < \ln(\phi_i) + (1 - \lambda)\phi_1' + \lambda\phi_2'; \\ 1, & \text{if } \ln(\phi_i - \eta) + (1 - \pi)\phi_1' + \pi\phi_2' \geq \ln(\phi_i) + (1 - \lambda)\phi_1' + \lambda\phi_2', \end{cases}$$

which reduces to

$$h = \begin{cases} 0 \text{ (don't invest)}, & \text{if } -\ln((\phi_i - \eta)/\phi_i) > (\pi - \lambda)(\phi_2' - \phi_1'); \\ 1 \text{ (invest)}, & \text{if } -\ln((\phi_i - \eta)/\phi_i) \leq (\pi - \lambda)(\phi_2' - \phi_1'). \end{cases}$$

The household will invest in their children if the decrease in utility from investing, $-\ln[(\phi_i - \eta)/\phi_i]$, is less than the expected utility gain from raising their children's productivity, $(\pi - \lambda)(\phi_2' - \phi_1')$.

It is easy to deduce that the analogous decision for a type-i single mother (for $i = 1, 2$) is represented by

$$h = \begin{cases} 0 \text{ (don't invest)}, & \text{if } -\ln((\phi_i - 2\eta)/\phi_i) > (\pi - \lambda)(\phi_2' - \phi_1'); \\ 1 \text{ (invest)}, & \text{if } -\ln((\phi_i - 2\eta)/\phi_i) \leq (\pi - \lambda)(\phi_2' - \phi_1'). \end{cases}$$

A single mother has half the income of a married household of the same type. Therefore, relative to household income, it costs twice as much for a single mother to educate her children.

7.4   Being Raised by a Single Mother

It is easy to construct a situation where;

(1) Some children will grow up with a single mother;

(2) Children who grow up with single mothers have on average lower levels of human capital than those who don't;

(3) Girls who grow up living with a single mother are more likely to become single mothers than girls who grow up in a two-parent family. This results in a poverty cycle.

To construct such an equilibrium, the following conditions are assumed:

Condition 2 (All married couples educate their children)   *This can be guaranteed by assuming*

$$-\ln\left(\frac{\phi_1 - \eta}{\phi_1}\right) \leq (\pi - \lambda)(\phi_2' - \phi_1').$$

Observe that if the above equation holds for a type-1 household, then it certainly holds for a type-2 one, because $-\ln(\phi_i - \eta)/\phi_i$ is decreasing in ϕ_i. The cost of educating children is less of a burden for the richer type-2 families.

Condition 3 (Single mothers do not educate their children)   *This can be achieved by imposing the restriction*

$$-\ln\left(\frac{\phi_2 - 2\eta}{\phi_2}\right) > (\pi - \lambda)(\phi_2' - \phi_1').$$

If a high-type single mother does not educate her child, it is easy to deduce that a low-type one won't either, because $-\ln(\phi_i - 2\eta)/\phi_i$ is decreasing in ϕ_i. When educating children is too expensive for the better-off type-2 mother, this will certainly be true for the poorer type-1 mother.

Condition 4 (Couples with a high bliss shock, b, always marry)   *To obtain this result, let*

$$\ln(\phi_1 - \eta) + (1 - \pi)\phi_1' + \pi\phi_2' + b \geq \ln(\phi_1/2) + (1 - \lambda)\phi_1' + \lambda\phi_2'.$$

The left-hand side gives the value of marriage for a type-1 couple that draws the high value for the bliss shock, b. The right-hand side gives the value of single life for the type-ϕ_1 woman. Recall that a single mother will not to educate her children. Also, note that value of single life for a type-1 man is $\ln(\phi_1/2)$. So, if the above equation holds, then a type-1 man will want to marry as well, since $\ln(\phi_1/2) + (1 - \lambda)\phi_1' + \lambda\phi_2' > \ln(\phi_1/2)$. That is, single life is worse for a type-1 man relative to a type-1 women because he will be estranged from his children. Again, note that $\ln[(\phi_i - \eta)/\phi_i]$ is increasing in ϕ_i so that if the above condition holds for ϕ_1, then it also holds for ϕ_2. Therefore, all couples that draw a high value for the bliss shock will marry. This transpires because the richer type-2 married households can more easily afford to educate their offspring.

Condition 5 (High-productivity couples with a bad bliss shock choose to marry while low-productivity ones do not) *Two assumptions are needed:*

$$\ln(\phi_2 - \eta) + (1 - \pi)\phi_2' + \pi\phi_2' - b \geq \ln(\phi_2/2) + (1 - \lambda)\phi_1' + \lambda\phi_2',$$

and

$$\ln(\phi_1 - \eta) + (1 - \pi)\phi_1' + \pi\phi_2' - b < \ln(\phi_1/2) + (1 - \lambda)\phi_1' + \lambda\phi_2'.$$

Here type-ϕ_2 households that draw a low value for the bliss shock, $-b$, will desire to marry. A type-ϕ_1 woman will prefer single life over married life when the bliss shock is low. Observe that the last condition can be rewritten as

$$b > \ln\left(\frac{2\phi_1 - 2\eta}{\phi_1}\right) + (\pi - \lambda)(\phi_2' - \phi_1').$$

This condition can be made to hold by picking a b that is large enough. If a type-ϕ_1 woman does not want to marry, then it does not matter what a type-ϕ_1 male prefers.

7.4.1 The steady-state fraction of women who are single mothers

Given the above conditions, type-ϕ_1 matches that draw the low value for bliss shock, $-b$, will not result in a marriage. The single woman will not educate her children. Let v denote the steady-state fraction of women who have low productivity. This is an endogenous variable. By assumption, v will not change over time. The steady-state fraction of women with the low level of productivity is determined by the formula

$$v = v\beta(1 - \lambda) + v(1 - \beta)(1 - \pi) + (1 - v)(1 - \pi).$$

To understand this equation, focus on the right-hand side. There are exactly three ways a girl can become a low-productivity women. To begin with, she may have had a low-productivity mother. There are v low-productivity mothers in the population. The fraction β of these will remain unmarried. Their children will be a low type with probability $1 - \lambda$. Thus, the fraction of women who are both low-type and who are born to low-type single mothers is $v\beta(1 - \lambda)$. This explains the first term. A low-productivity mother may marry with probability $1 - \beta$. Even though the children in her household will be educated, they may still turn out to be a low type with probability $1 - \pi$. This gives the second term. Third, there are $1 - v$ high-type marriages in the population. A fraction $1 - \pi$ of these marriages will have low-type children. Therefore, the number of low-type girls arising from high-type marriages is $(1 - v)(1 - \pi)$. Consequently, the right-hand side gives the fraction of women who are low-type, which in a steady state must equal v, or the left-hand side. Solving for v gives

$$0 < v = \frac{1 - \pi}{1 - \beta[(1 - \lambda) - (1 - \pi)]} < 1.$$

Since the fraction of low-type women who don't marry is β, the fraction of women who are single mothers is

$$\beta v = \frac{\beta(1 - \pi)}{1 - \beta[(1 - \lambda) - (1 - \pi)]}.$$

To summarize:

(1) The fraction of single mothers among women is $0 < v\beta < 1$. Not surprisingly, the proportion of single mothers is increasing with the odds that a low-type mother will not get married, β. It also rises in the probabilities that children will turn out to be a low type, $1 - \lambda$ and $1 - \pi$.

(2) The expected level of human capital for a child growing up with a single mother is $(1 - \lambda)\phi_1'/2 + \lambda\phi_2'/2$, which is less than the expected level of human capital for a child growing up in a two-parent family, $(1 - \pi)\phi_1'/2 + \pi\phi_2'/2$.

(3) The odds of a girl who grew with a single mother becoming a single mother herself are $\beta(1 - \lambda)$, while the probability of a girl becoming a single mother who grew up in a two-parent family is $\beta(1 - \pi)$. Therefore, girls growing up with a single mother are more likely to become single mothers themselves than girls growing up with two parents, because $\beta(1 - \lambda) > \beta(1 - \pi)$. Single parenthood begets single parenthood and a poverty cycle results.

8 The Beckerian Theory of Marriage

Becker (1973) developed a famous theory of marriage. He argued that individuals on the marriage market will sort so that couples are matched together efficiently. Efficient matching maximizes the aggregate output of "marital goods" that households produce. Additionally, efficient matching implies that it is impossible to rearrange marriages so that some pairs of individuals can both be made better off by matching them together. Becker (1973) also argued that there will be positive assortative mating, at least if productive men complement productive women in producing marital output. That is, highly (lowly) valued individuals will marry other highly (lowly) valued individuals. The Becker marriage model applies, in theory, also to cohabiting couples and same-sex marriages.

8.1   Stable and Efficient Matching

Suppose that a person has a simple linear utility function of the form

$$U(c) = c,$$

where c represents the consumption of household goods. Superscripts m and f will be attached to c to denote consumptions by men and women. A person cares only about her or his personal consumption, which is a private good. Becker (1973) assumes that within a marriage it is possible for one person to transfer money to the other, which will decrease the private consumption of the former and increase the consumption of the latter. This is in contrast with the previous model of marriage where the utility generated by consumption was a public good collectively enjoyed by each partner in the exact same manner. When all consumptions are public goods, there is no opportunity for one party to transfer consumption to the other. Let there be N types of men and women indexed by i and j respectively. There are an equal number (normalized to one) of people of each type. (For same-sex marriages just divide the population into two identical halves. To map things into the analysis below, call one half men and the other half women even though both parties are of the same sex.) Suppose that a man and woman of types i and j can produce marital output in the amount $o_{i,j}$. There are N^2 possible types of marriages.

Think about all the possible ways that men and women could be matched together. You could match the first type of man with N types of women. This leaves $N-1$ types of women left for the second type of man. Once he is matched, the third type of man can be paired with any of $N-2$ types of women, and so on. Thus, there are $N!$ possible matching schemes. Denote a particular matching scheme by the rule $j = M(i)$, for $i = 1, \cdots, N$, which maps a man of type i to a woman of type j. Thus, M lies in a set $\mathcal{M}$ that contains $N!$ such rules.

Let

$$O^M \equiv \sum_i^N o_{i,M(i)}$$

represent the aggregate level of marital output that is associated with matching scheme M. The output maximizing matching scheme, M^*, is given by

$$M^* = \arg\max_{M \in \mathcal{M}}\{O^M\}.$$

By definition this matching scheme maximizes aggregate marital output. Call the matching rule efficient. Now, relabel men so that a type-i man matches with a type-i women under the efficient matching scheme. That is, reorder men so that the efficient matching rule gives $i = M^*(i)$. Then,

$$O^{M^*} \equiv \sum_{i=1}^{N} o_{i,i} = \sum_{i}^{N} o_{i,M^*(i)}.$$

Consider some proposed matching scheme M, which need not maximize aggregate marital output. The sum of consumptions by the man and woman in the household must satisfy

$$c_i^m + c_{M(i)}^f = o_{i,M(i)},$$

where c_i^m and $c_{M(i)}^f$ represent the levels of consumption under this matching scheme for a man of type i and the woman who is matched with this man, $M(i)$. Recall that consumption, and hence utility, is completely transferable between the man and woman in a marriage. For this matching scheme to be viable, it must be the case that it is not in the interest of *any* man or woman, say i or j, to peel off and form his or her own match with someone else outside of this scheme. Denote the consumption of a man and woman of *arbitrary* types, i and j, under the matching scheme M by c_i^m and c_j^f. This pair will not, in general, be matched to each other under the proposed matching scheme. The following defines the notation of a stable matching scheme:

Definition 2 (Stable Matching Scheme) *A matching scheme is stable if $c_i^m + c_j^f \geq o_{i,j}$, for all i, j, where c_i^m and c_j^f are levels of consumptions for man i and women j under the matching scheme and o_{ij} is the level of marital output that obtains if i and j were matched together.*

That is, it cannot be possible that by pairing i and j together, they could gain a higher level of total consumption, $o_{i,j}$, than they are currently obtaining, $c_i^m + c_j^f$, under matching scheme M. If this is possible, then i and j could form a marriage, producing some level of output, $o_{i,j}$, which yields one or both of them a higher level of consumption. A stable matching scheme is therefore Pareto optimal. In other words, it is not possible to rearrange the matches so as to make one person better off without making someone else worse off. If it is possible to rearrange the matches so as to make one person better off without making someone else worse off, then the matching scheme is not stable.

Proposition 1 (Stable Matching, Sufficiency)   *A matching scheme maximizes aggregate marital output if it is stable.*

Proof   Take any stable matching scheme, $j = M'(i)$, in the feasible set, $\mathcal{M}$. Represent the consumptions for individuals i and j that are associated with this stable matching scheme by $c_i^{m'}$ and $c_j^{f'}$. Next, among all matching schemes $j = M(i)$ in the feasible set $\mathcal{M}$, pick the matching scheme $j = M^*(i)$ that maximizes the aggregate level of output. Thus, the efficient matching rule M^* is specified by

$$M^* = \arg\max_{M \in \mathcal{M}} \sum_i^N o_{i,M(i)}.$$

By the definition of a stable matching scheme, it must transpire that

$$c_i^{m'} + c_{M'(i)}^{f'} \geq o_{i,M^*(i)}.$$

That is, under the stable matching scheme the sum of the consumptions that individuals i and $j = M'(i)$ get must be at least as great as what they can produce together under the scheme that maximizes aggregate output, $j = M^*(i)$. Otherwise, they could split off from the stable scheme and join together to produce $o_{i,M^*(i)}$, which would result in at least one or both of them attaining a higher level of consumption. Summing over all i's then gives

$$O^{M'} \equiv \sum_i^N c_i^{m'} + c_{M'(i)}^{f'} \geq \sum_i^N o_{i,M^*(i)} \equiv O^{M^*}.$$

This implies that aggregate output under the stable matching scheme, $O^{M'}$, must be at least as great as aggregate output, O^{M^*}, under the scheme $j = M^*(i)$ that maximizes aggregate output. Thus, the stable matching scheme M' gives the highest level of marital output in the set of all possible matching schemes, $\mathcal{M}$. ∎

Given that partner sorting à la Becker maximizes aggregate marital output, the efficient matching scheme must solve the following linear programming problem:

$$\max_{\{m_{ij}\geq 0\}} \sum_i^N \sum_j^N m_{ij}o_{ij},$$

subject to

$$\sum_j^N m_{ij} = 1, \quad \text{for } i=1,\cdots,N,$$

and

$$\sum_i^N m_{ij} = 1, \quad \text{for } j=1,\cdots,N.$$

The choice variables m_{ij} (for all i and j) give the number of matches between a man of type i and a woman of type j. The first constraint ensures that the number of matches for type-i men cannot exceed their mass in the population. Such a constraint holds for each $i=1,\cdots,N$. Note that these constraints, together with the restriction $m_{ij} \geq 0$, imply that $m_{ij} \leq 1$. The second constraint does the same thing for a type-j women.

Let c_i^m denote the Lagrange multiplier associated with the first constraint and c_j^f be the multiplier connected with the second (the Lagrange multiplier is covered in the Mathematical Appendix).[7] The first-order condition for m_{ij}

7. Set up the Lagrangian function, $L(m_{ij}, c_i^m, c_j^f)$, as follows:

$$L(m_{ij}, c_i^m, c_j^f) = \sum_i^N \sum_j^N m_{ij}o_{ij} + \sum_i^N c_i^m (s_i^m - \sum_j m_{ij}) + \sum_j^N c_j^f (s_j^f - \sum_i m_{ij}).$$

where $s_i^m = 1$ and $s_j^f = 1$ are the supplies of type-i men and type-j women. Taking the derivatives with respect to m_{ij}, c_i^m, and c_j^f yields the first-order conditions shown below.

$$\frac{dL(m_{ij}, c_i^m, c_j^f)}{dm_{ij}} = o_{ij} - c_i^m - c_j^f \leq 0, \quad \text{for all } i,j=1,\cdots,N,$$

$$\frac{dL(m_{ij}, c_i^m, c_j^f)}{dc_i^m} = s_i^m - \sum_j^N m_{ij} = 0, \quad \text{for } i=1,\cdots,N,$$

in the above maximization problem is

$$o_{ij} \leq c_i^m + c_j^f, \text{ for all } i, j = 1, \cdots, N.$$

Here c_i^m is the shadow value of a type-i man in the pool of males. This is the price of a type-i man, so to speak. Likewise, c_j^f is the price of a type-j woman. When the above first-order condition is slack, so that $o_{ij} < c_i^m + c_j^f$, then $m_{ij} = 0$. This says that if a match between a type-i man and a type-j women does not occur, then the value of the match, o_{ij}, falls below the price of the type-i man plus the price of the type-j woman. Now, when $m_{ij} > 0$, then $c_i^m + c_j^f = o_{ij}$. The man's share of the marital output, o_{ij}, is c_i^m, while the woman receives c_j^f. Interestingly, the above first-order condition is also the condition for a stable match. This leads to the proposition below.

Proposition 2 (Stable Matching, Necessity) *A matching scheme maximizes aggregate marital output only if it is stable.*

Example 2 (Efficient matching—Becker, 1973, p. 824) Consider table 3.7, which shows the potential marriages between men and women that can occur when there are just two types for each sex.

The efficient matching rule pairs a type-i man with a type-i woman so that $i = M^*(i)$, for $i = 1, 2$; i.e., matches along the diagonal are the efficient ones. Aggregate marital output is $O^{M^*} = 8 + 7 = 15$. The maximum marital output among individual marriages occurs when $i = 2$ and $j = 1$. So why doesn't the type-2 man

---

and

$$\frac{dL(m_{ij}, c_i^m, c_j^f)}{dc_j^f} = s_j^f - \sum_i^N m_{ij} = 0, \quad \text{for } j = 1, \cdots, N.$$

Observe that the last two first-order conditions are just the constraints. The solution to this system of first-order condition describes the solution to the maximation problem.

An application of the envelope theorem to the *optimized* function $L(m_{ij}, c_i^m, c_j^f)$ also gives

$$\frac{dL(m_{ij}, c_i^m, c_j^f)}{ds_i^m} = c_i^m \quad \text{and} \quad \frac{dL(m_{ij}, c_i^m, c_j^f)}{ds_j^f} = c_j^f.$$

This implies that the values of extra type-i men and type-j women are c_i^m and c_j^f, respectively.

Table 3.7
Matching.

Men	Women	
	$j=1$	$j=2$
$i=1$	$o_{1,1}=8$	$o_{1,2}=4$
$i=2$	$o_{2,1}=9$	$o_{2,2}=7$

Source: Becker 1973.

match with the type-1 woman? This would imply that a type-1 man would have to match with a type-2 woman. The aggregate marital output from this match is $O^M = 9 + 4 = 13 < O^{M*}$. Under the efficient marriage scheme it is possible to split up the output from the marriages as follows: $c_1^{m*} = 3, c_1^{f*} = 8 - 3 = 5, c_2^{m*} = 5,$ and $c_2^{f*} = 7 - 5 = 2$. Observe that $c_2^{m*} + c_1^{f*} = 5 + 5 = 10 > 9 = o_{2,1}$. Therefore, it is impossible for a type-$(2, 1)$ marriage to provide this much total consumption, which could be split up between husband and wife. (Likewise, note that a type-$(1, 2)$ marriage would deliver less total consumption than the sum of consumptions that a type-1 man and type-2 woman get under the efficient scheme.) This illustrates that an efficient matching rule results in stable matching, wherein it is impossible to rearrange the marriages to make some couples better off. A type-2 man and a type-1 woman receive a lot when marriages are efficient, but their partners receive less. This follows from the transferable utility assumption. This allows each of them to do better in their respective marriages than they would in the high output type-$(2, 1)$ marriage.

8.2 Assortative Mating

Let a higher index for either a man or woman denote a better type and presume that marital output is increasing in the types of the partners. Suppose that the function $o_{i,j}$ exhibits Edgeworth-Pareto complementarity (see chapter 1, definition 3). Specifically, assume that

$$\frac{\partial^2 o_{i,j}}{\partial i \partial j} > 0. \tag{8.1}$$

One can think about $\partial o_{i,j}/\partial i$ as giving the marginal product of a man's type in a marriage. Therefore, the condition $\partial^2 o_{i,j}/(\partial i \partial j) > 0$ states that the marginal product of a man's type in a marriage is increasing in the woman's type j. Similarly, by Young's theorem, the marginal product of a woman's type, j, is increasing in the man's type i. Condition (8.1) implies that

$$\int_j^l \frac{\partial o_{i,x}}{\partial x} dx > \int_j^l \frac{\partial o_{i-1,x}}{\partial x} dx > \cdots > \int_j^l \frac{\partial o_{k,x}}{\partial x} dx, \text{ for } l > j \text{ and } k < i-1,$$

because $\partial o_{i,x}/\partial x > \partial o_{i-1,x}/\partial x > \partial o_{k,x}/\partial x$. But, from the fundamental theorem of calculus, $o_{i,l} - o_{i,j} = \int_j^l (\partial o_{i,x}/\partial x) dx$ for all i. Therefore,

$$o_{i,l} - o_{i,j} > o_{i-1,l} - o_{i-1,j} > \cdots > o_{k,l} - o_{k,j}, \quad \text{for } l > j \text{ and } k < i-1.$$

Simply put, the difference in marital outputs between women of types l and j is increasing in the male's type, i. This property is called *increasing differences*.

Proposition 3 (Positive Assortative Mating) *The efficient matching rule is* $i = M^*(i)$.

Proof Consider some other matching rule, $j = M(i)$. This rule cannot maximize output. Make a list from the best man down to the worst one; i.e., the indices in the list will be descending. Find the first instance where $j = M(i)$ with $j \neq i$. Note that it must be the case that $i > j$. Locate the male, k, who is assigned with woman i or find the k that gives $i = M(k)$. Now rematch man i with woman i to get an increase in output of $o_{i,i} - o_{i,j} > 0$, because $i > j$. This leaves woman j and man k without mates. Next, reassign woman j with man k to get a gain in output of the amount $-(o_{k,i} - o_{k,j}) < 0$, because $i > j$. Finally, observe that $o_{i,i} - o_{i,j} > o_{k,i} - o_{k,j}$, since $i > k$. (To see this, set $l = i$ in the above increasing differences formula and then compare the first and last terms). ∎

The upshot of this section is that Becker's theory of marital sorting implies that

(1) Marriages will be arranged in an efficient manner in the sense that they maximize aggregate marital output and that it is impossible to rearrange matches so that some reformed couples are better off;

(2) There will be positive assorting mating whereby high types will marry other high types and low types will marry other low types, at least under the assumption that female and male types are Edgeworth-Pareto complements in marital production.

9 Literature Review

Ogburn and Nimkoff (1955) noted the changing nature of marriage in a very prescient book. Interest in economic theories of marriage took off with a famous paper by Becker (1973). A basic model of the rise in divorce and the decline in marriage is provided in Greenwood and Guner (2009). They stress labor-saving technological progress in the home and growth in wages as the driving forces. In an extension, they also model the decision for a young adult to leave the parents' home, either as a single or married person. The rising standards of living that made it possible for young people to live alone have also made it possible for elderly widows to do the same, according to Bethencourt and Rios-Rull (2009). Santos and Weiss (2016) observe that children are a large expense that parents cannot easily avoid during tough economic times. Accordingly, they argue that people delay families, and thus marriage, when there is instability in the labor market. Instability in the labor market rose in last quarter of the twentieth century and has continued into this one. Fernandez and Wong (2017) examine the implications of a move towards unilateral divorce in the 1970s. They find that this switch promoted divorce and increased married female labor-force participation. In the current analysis, think about this as a drop in λ. Fernandez and Wong (2017) also find that a reduction in the gender gap barely changed divorce and led to a small reduction in the number of married women. It did have a significant impact on married female labor-force participation. The impact of such laws on savings is studied in Voena (2015). She finds that the unilateral divorce combined with an equal division of property encourages savings and discourages female labor-force participation.

The Gale-Shapley (1962) matching algorithm is detailed in the book by Browning, Chiappori, and Weiss (2014). The recent rise in assortative mating is analyzed in Greenwood, Guner, Kocharkov, and Santos (2016). They find that shifts in the U.S. wage structure favoring skilled labor has had a powerful influence on assortative mating. Couples may sort on many dimensions. Chiappori, Oreffice, and Quintana-Domeque (2012) show that body mass is one such factor. Positive assortative mating may provide a marriage-market return for educational investment, in addition to the traditional labor-market return. By getting an education, one can attract a better partner, or in the terms of the discussion in this chapter, move up on the matching table. This is analyzed for women in Chiappori, Iyigun, and Weiss (2009).

McLanahan and Sandefur (1994) documented the status of children growing up with a single mother. Appendix A in their book details the databases and variables used. Aiyagari, Greenwood, and Guner (2000) and Gayle, Golan, and Soytas (2015) develop economic models of this issue. Regalia and Rios-Rull (2001) use a model of marriage and divorce to account for the rise in the number of single mothers. They stress market forces, such as a movement in the gender gap, as explaining this rise. Caucutt, Guner, and Rauh (2017) analyze the difference in marriage rates for blacks and whites in the United States. They suggest that current incarceration policies and labor market prospects make black men much riskier spouses than white men.

Both Browning, Chiappori, and Weiss (2014) and Ermisch (2003) provide detailed presentations of economic models of marriage and divorce, as well as different modes of interaction (cooperative and noncooperative) between husbands and wives. The facts on cohabitation for opposite- and same-sex relationships are taken from Blau, Ferber, and Winkler (2014). Gemici and Laufer (2010) study opposite-sex cohabitation. They find that relative to traditional marriage, opposite-sex cohabitation is connected with less household specialization, greater instability in relationships, and stronger positive assortative mating.

10 Problems

(1) *The changing value of women's work.* Take the set-up that is presented for perfectly assortative mating in Section 6.1. Now, let $\theta = 0.3$, $\kappa = 0.7, c = 0, d = 1, q = 0.2, w = 4$, $\mu_1 = 0.5, \mu_2 = 1.0$, $\phi_1 = 0.5, \phi_2 = 0.6$, $\eta_1 = 0.5, \eta_2 = 3.0, \pi_1 = \pi_2 = 0.5$, and $\chi_{11} = \chi_{12} = \chi_{21} = \chi_{22} = 0.25$. What will the matching table look like? (Watch out for a corner solution.) Change $d = 2.0$ and $\kappa = 0.05$. Now, what does the matching table look like? Explain the economics behind this.

(2) *Single life: a lack of options and/or a choice?* Imagine a world with both men and women in which males are all the same, but there are two types of women. These women differ by their productivity in the market, ϕ_j, and at home, η_j, for $j = 1, 2$, with $\phi_2 > \phi_1$ and $\eta_2 > \eta_1$. The fraction of women of each type is given by χ_j. A male's productivity at home and in the market

is 1. Each sex has one unit of time. An efficiency unit of labor supplied to the market earns the wage rate w; i.e., if a person's productivity in the market is π, then they will earn $w\pi$ for each unit of labor supplied. Suppose that tastes for a single person are given by

$$\theta c + (1 - \theta) \ln(n), \text{ with } 0 < \theta < 1,$$

where c is the individual's consumption of market goods and n is the consumption of home goods. A single person produces household goods according to the home production function

$$n = d^{1-\kappa}(\lambda l)^{\kappa}, \text{ with } 0 < \kappa < 1,$$

where l is the household's labor at home, d represents the inputs of goods into home production, and λ is the person's productivity at home. Suppose that d is fixed and can be purchased at the price q. (Assume that $w\phi_1 > q$.) A married household has tastes represented by

$$\theta c/2 + (1 - \theta) \ln(n/2^k), \text{ again with } 0 < \theta < 1,$$

where c is the couple's consumption of market goods and n is their consumption of home goods. In a married household, only the woman engages in household production. She is free to split her time between market and nonmarket work. The married household uses the same production function for home goods as the single one. Suppose that all matching is perfectly assortative. (Assume an interior solution below.)

(a) Analyze marriage in this economy. Is it possible that some people may not want to marry? Discuss when this is more likely to occur.

(b) Suppose that the government taxes married households in the lump-sum amount t and subsidizes single households in the amount $s/2$. How might this affect things? Why is this thought experiment important?

4   Social Change

Why is there so much social change today, and why was there so little in ancient times? The most probable answer, the result of quite extensive study, is mechanical invention and scientific discovery. There is no doubt that useful inventions and researches cause social changes. Steam and steel were major forces in developing our extensive urban life. Gunpowder influenced the decline of feudalism. The discovery of seed-planting destroyed the hunting cultures and brought a radically new form of social life. The automobile is helping to create the metropolitan community. Small inventions, likewise, have far-reaching effects. The coin-in-the-slot device changes the range and nature of salesmanship, radically affects different businesses, and creates unemployment. The effects of the invention of contraceptives on population and social institutions is so vast as to defy human estimation. It is obvious, then, that social changes are caused by inventions.

—William F. Ogburn (1936)

Economics typically either ignores culture or takes it as an exogenous factor in the economy. The famous sociologist William F. Ogburn thought that social change sprung from technological progress, albeit sometimes with a long lag. This chapter models social change as a function of technological advance. It shows how shifts in women's rights in the workplace, changes in attitudes about married women working, the sexual revolution, and evolution in ideas about the virtues of patience and education may all be in response to technological developments in the economic environment.

The formal analysis begins with a model of the evolution of women's rights in the work place. This is done with the context of the median voter model. This is an abstraction, obviously. There was no formal vote to grant women rights in the workplace. So, it is shorthand for saying that as a greater fraction of the population became more accepting of women working, women's rights in the workplace expanded through favorable court decisions, the passage of laws, and changes in business and government employment practices. Such change happened relatively quickly starting in the 1960s. Thus, modeling it as happening in a single period of time is not a great violation. The issue formulated is whether or not society will vote to grant married women the right to work. There is a tradeoff involved. On the one hand, by granting women the right to work, households will earn more income. On the other hand, it takes some of a mother's time away from home. There may be a cost to this, if it reduces the quality of society's children, as reflected by their knowledge and cognitive and noncognitive skills. In addition to prizing the quality of their own children, a household values the quality of other people's kids. It is shown that both technological progress in the market, which raises wages, and technological advance in household sector, which frees up a woman's time at home, are conducive to granting married women the right to work. Labor-saving technological advance in the household sector also means that women can go to work *without* spending less time with their kids. This is the first example of technological advance causing social change.

The next section models the evolution of attitudes toward married women working. Here it is assumed that a household is more receptive to a wife working if the husband's mother worked. The analysis starts off from a situation where no married women work. A rise in wages entices some women to enter the labor force. This leads to the sons of these mothers being more welcoming to their own wives working. Hence, in the next generation, there will be even more working women. The process continues, with ever increasing labor-force participation rates by women until it finally converges to a new long-run equilibrium where working mothers are the norm. In this situation, economics affects culture, but culture also feeds back on economics.

The sexual revolution provides an ideal case study of social change. To some the phrase "sexual revolution" may conjure up visions of hippies in the 1960s, but many believe that the first sexual revolution actually happened in the 1920s. As will be seen, sexual mores and practices became more liberal throughout the twentieth century as contraceptive technology improved and knowledge about its use was disseminated. It is easy to identify improvements in the efficacy and use of contraception. The resulting changes in sexual behavior are also readily observable. How can the risk of engaging in premarital sex be measured? The main analysis starts off by discussing the improvements in contraceptive practices that happened over the course of the twentieth century. It is demonstrated that the odds of a sexually active teenage girl becoming pregnant declined rapidly over this period. As the above quotation illustrates, the farsighted Ogburn saw that contraception was becoming more effective well before the invention of the pill and that this technological advance would dramatically change society. The next section proceeds by modeling a teenage girl's decision to engage in premarital sex. In doing so, she rationally weighs the costs and benefits of sexual activity. It is shown how improvements in contraceptive technology will lead to a greater fraction of teenage girls becoming sexually active. This leads to a ∩-shaped pattern over time in the fraction of girls who have an out-of-wedlock birth. It may seem strange that the initial improvements in contraception lead to a rise in out-of-wedlock births—the left-hand side of the ∩—but this will be explained.

Parents can influence their children's behavior by influencing how their offspring weigh the costs and benefits from doing certain things. In times past parents dissuaded premarital sexual activity by shaming it. Technological progress in contraception has reduced the need to do so, but even today there are vestiges of shaming. Statistically speaking, the greater the shame a teenage girl will feel from becoming pregnant and/or the more religious she is, the less likely she will be sexually active. The next section discusses how parents can instill such shame in their daughter. The idea is that parents recognize that their teenage daughter will do what she pleases when she is beyond their direct oversight. By instilling a sense of shame in a daughter, should she have an out-of-wedlock birth, they can influence the daughter's cost/benefit calculation regarding premarital sex. The need to inculcate shame, however, diminishes with improvements in contraceptive technology. Shaming still plays a role today in discouraging certain behaviors, such as dishonesty, sloth, substance abuse, and violence. It is also discussed how encouraging a sense of pride for "doing the right thing" works in essentially the same way.

Historically churches and governments also had an interest in reducing out-of-wedlock births. They were responsible for providing charity and social assistance to unwed mothers, a major expense in the poor economies of the past. They clamped down on premarital sex in draconian ways. Shaming was one of the key ways they used to dissuade premarital sex. This was the reflected in the language they used to characterize illicit sex. So, for example, the son and namesake of the renowned minister John Cotton was excommunicated in 1664 by the First Church of Boston "for lascivious unclean practices with three women" (Godbeer 2002). It is easy to extend the model developed in the previous sections to include religious proscription against out-of-wedlock births. This is operationalized here by assuming that the church can have an influence on a parent's decision about how much to socialize a daughter. It is shown that as contraception becomes more effective, the extent of religious proscription will drop, as it indeed has. The big picture here is that the economy, via technology, influences culture, and culture in turn affects the economy by impinging on people's behavior.

Parents may also socialize their children about other things, such as the need to delay gratification today for gain tomorrow. That is, they may teach their children the virtues of patience. This is studied in the penultimate section of the chapter, which demonstrates how, when the value of an education rises, say, due to an industrial revolution that favors skilled labor, parents will feel the need to instill more patience in their offspring so that the latter will acquire education. Again, this is another illustration of how values in a society may change as the economy evolves. As always, the chapter concludes with a brief discussion of the literature upon which it draws.

1 Women's Rights in the Workplace

Some selected milestones for women's rights in the United States are presented in table 4.1. This list is far from complete. It emphasizes topics discussed here, to wit: married female labor-force participation, marriage and divorce, and reproduction. In 1920 an amendment to the U.S. Constitution granted women the right to vote. In the same year, the U.S. Department of Labor established the Women's Bureau. Its purpose was to collect information about women in the workforce and to improve their working conditions. The Equal Pay Act, passed in 1963, made it illegal to pay a woman less than what a man would earn for the same job. Around the same time, the Civil Rights Act, Title VII, prohibited employment discrimination on the basis of sex. The act established the Equal Employment Opportunity Commission (EEOC), which is charged with investigating complaints and imposing penalties. In 1970, in *Shultz v. Wheaton Glass Co.*, the U.S. Court of Appeals for the Third Circuit ruled that Title VII

Table 4.1
Some selected U.S. women's rights milestones.

Year	Event
1920	19th Amendment
1920	Women's Bureau
1936	*U.S. v. One Package*
1960	Birth control pill
1963	Equal Pay Act
1964	Civil Rights Act, Title VII
1964	Equal Employment Opportunity Commission
1965	*Griswold v. Connecticut*
1969	No Fault Divorce, California
1970	*Shultz v. Wheaton Glass Co.*
1972	Title IX, Education Amendments
1973	*Roe v. Wade*
1978	Pregnancy Discrimination Act
1986	*Meritor Savings Bank v. Vinson*

applied to jobs with "substantially equal" task requirements for men and women, although not necessarily in title or job description. The Wheaton Glass Company had employed male "selector-packer-stackers" and female "selector-packers." Although they performed similar work in the company's warehouse, the selector-packer-stackers' job classification paid materially more than the selector-packers' one. Title IX of the Education Amendments, passed in 1972, banned discrimination against women in education. This facilitated entry into professional schools, among other things. Employment discrimination against pregnant women was prohibited in 1978. Another issue was sexual harassment. After being fired from her job at Meritor Savings Bank, Mechelle Vinson sued Sidney Taylor, the vice president of the bank. Vinson said that Taylor coerced her into have sexual relations with him 40 or 50 times. Her suit stated that the sexual harassment created a hostile working environment, which amounted to unlawful discrimination under Title VII of the Civil Rights Act of 1964. In 1986 the Supreme Court agreed and found that sexual harassment on the job violated Title VII of the Civil Rights Act.

The other entries in the table refer to marriage and reproductive rights. No-fault divorce made it easier for both men and women to terminate marriages. Wives were no longer tethered to their husbands. This was touched upon in chapter 3. The entries on *U.S. v. One Package*, the birth control pill, *Griswold v. Connecticut*, and *Roe v. Wade* concern reproductive rights. These are detailed later in this chapter.

1.1   Setting up the Model of Women's Rights

A stylized model of the process whereby women gain rights in the workplace is now formulated. The vehicle for analysis is the median voter model. Imagine an economy populated by married households, each with two children. Both the husband and wife have one unit of time. The husband spends all of his time working in the market at the wage w. The wife has three potential uses of her time: working at home, spending time with her children, and working in the market at the wage rate rw, where $r \in \{0, \phi\}$ reflects women's rights in the workplace. When $r = 0$ (no rights) married women are prohibited from working, while when $r = \phi$ (equal rights) they can work at the wage ϕw.[1] Here, ϕ represents the gender wage gap when women have rights in the workplace. In this stylized world without discrimination, women are still paid differently than men at a particular point in time. This could occur because jobs in the past required more brawn than brain—see chapter 1—or because women worked in lower-paying occupations that were more accommodating about taking time off for children—see chapter 2. In this stylized world, if there is discrimination, then the gender wage gap is 0; i.e., women can earn nothing in the labor force. So, discrimination obviously affects the gender wage gap. Labor is indivisible. A mother must spend the *fixed* amount of time $\mathfrak{h}$ on housework. If the woman works in the market, then she must work the *fixed* amount $\mathfrak{l}$ there.

The household's utility function is given by

$$c + \lambda q + \eta \mathbf{q},$$

where c is the household's consumption, q is the quality of their two children, and $\mathbf{q}$ is the *average* quality of children in society. The quality of children is captured by their knowledge gained through formal and

---

1. Think about the situation corresponding to $r = 0$ as reflecting a world where there is severe discrimination against women, which makes it undesirable for them to work. This equilibrium could be supported by a trigger strategy mechanism, whereby people won't truck with those who hire women or those who do business with those who hire women.

informal education. It also includes cognative and noncognative skills. The constant η measures how households in society care about other people's children, as opposed to their own,which is reflected by λ. The variable λ is distributed across households according to a uniform distribution on the interval $[0, 1]$. (The uniform distribution is defined in the Mathematical Appendix.) Thus, some households care more about the quality of their offspring (relative to consumption) than others. The quality of children is specified by

$$q = \ln t,$$

where t is the mother's time spent with them. Households also care about the average quality of children in society, $\mathbf{q}$. Low-quality children in society may lead to social problems such as crime, unwanted pregnancies, and unemployment.

The idea for this model, in a nutshell, is that when a household decides how to vote, it weighs the extra income a women can bring home against their concern that more working women may be bad for society's children. As the general level of wages rise, the gender gap shrinks, and technological progress reduces the need for labor at home, it becomes more advantageous for married women to work relative to staying at home. More and more families find it in their own best interest for the wife to participate in the labor force. This leads to a shift in attitudes about married women working. Eventually, a majority of families will be in favor of women working. In fact, with labor-saving technological progress in the household sector, it may be possible for women to go to work without spending less time with their children. Hence, the quality of children need not decline as more women work. As a consequence of these shifts in the economy, society votes to grant married women the right to work. The analysis now proceeds in three steps. First, a household's decision about whether the woman should work is posed and solved. Second, the median household's voting decision is worked out. Third, the model's implications for the evolution of women's rights are discussed.

1.2 The Decision to Work or Not

The household must decide whether or not the woman should work. The woman's unit of time is split between housework, $\mathfrak{h}$, working in the market, $l \in \{0, 1\}$, and improving the quality of the household's children, t; thus,

$t = 1 - \mathfrak{h} - l$. The household's decision regarding the woman's labor supply amounts to solving the maximization problem

$$\max_{l \in \{0,\mathfrak{l}\}} \{w + rwl + \lambda \ln(1 - \mathfrak{h} - l) + \eta \mathbf{q}\},$$

where l indicates whether the wife works in the market ($l = \mathfrak{l}$) or not ($l = 0$). This is a version of the married female labor supply model along the extensive margin that is presented in chapter 1. On the one hand, if the woman works, the household's utility will be $w + rwl + \lambda \ln(1 - \mathfrak{h} - \mathfrak{l}) + \eta \mathbf{q}$. On other hand, if she doesn't, then it is $w + \lambda \ln(1 - \mathfrak{h}) + \eta \mathbf{q}$. So, the decision to work can be represented as

$$l = \begin{cases} \mathfrak{l} \text{ (work),} & \text{if } rw\mathfrak{l} + \lambda \ln(1 - \mathfrak{h} - \mathfrak{l}) + \eta \mathbf{q} \geq \lambda \ln(1 - \mathfrak{h}) + \eta \mathbf{q}; \\ 0 \text{ (don't work),} & \text{if } rw\mathfrak{l} + \lambda \ln(1 - \mathfrak{h} - \mathfrak{l}) + \eta \mathbf{q} < \lambda \ln(1 - \mathfrak{h}) + \eta \mathbf{q}. \end{cases}$$

(Note that in the comparison, the w term will cancel out from both sides of the expression.) Define λ^* as the threshold value for λ where a household is indifferent with regard to whether or not the woman works in the market. It is easy to deduce that

$$\lambda^* = \Lambda(r) \equiv rw\mathfrak{l} / \ln[(1 - \mathfrak{h})/(1 - \mathfrak{l} - \mathfrak{h})]. \tag{1.36}$$

The decision for the woman to work can now be expressed as

$$l = \begin{cases} \mathfrak{l} \text{ (work),} & \text{if } \lambda \leq \lambda^* = \Lambda(r); \\ 0 \text{ (don't work),} & \text{otherwise,} \end{cases}$$

where λ^* is given by (1.36). In this setting, households that place a high value on the quality of the children (as represented by a value for λ greater than λ^*) will have a mother who stays at home. The fraction of women working in society is just

$$\Pr[\lambda \leq \lambda^*] = \lambda^* = \Lambda(r),$$

which follows from the fact that λ is uniformly distributed on $[0, 1]$. The extent of married female labor-force participation is a function of women's rights. It is clear that when $r = 0$ (an absence of rights), then $\lambda^* = \Lambda(0) = 0$, and no women will work.

Lemma 1 (Married female labor-force participation) *The fraction of women working in society, $\lambda^* = \Lambda(r)$, is increasing with women's rights, r, and the market wage rate, w, and is decreasing with the amount of housework required, $\mathfrak{h}$. Married female labor-force participation is not affected by the average quality of children in society, $\mathbf{q}$.*

Observe that the average quality of children in society, q, does not influence whether or not the woman in a household works in the market, at least given the status of women's rights, r. Why should it? The household cannot influence q, which essentially represents an externality and it enters the utility function in an additive way. It turns out that the status of women's rights is influenced by the quality of children, though, and vice versa. As will be seen, this is the channel through which the economic environment affects women's rights.

1.3 The Vote to Grant Rights or Not

Marriage bars to employment existed in the first half of the twentieth century. These were regulations that prevented single women from working after marriage and that prohibited employers from hiring married ones. In 1928, for example, 62 percent of school districts would not hire married women. This figure rose to 77 percent in 1942. Private firms had similar proscriptions.

Suppose that households can vote on a law that allows married women to work or not. When $r = \phi$, women can work, and when $r = 0$, they cannot. For a law to pass, 50 percent of households must vote for it. The median voter in society is the household with $\lambda = 0.5$. The law will pass if and only if the median voter household is in favor of it. This household plays an important role in what follows. Now, the average quality of children in society will be a function of whether or not women have the right to work. When $r = 0$, no woman will work. Then the average quality of children will be $q = \ln(1 - \mathfrak{h})$. When $r = \phi$, the fraction of married households with a working wife is $\lambda^* = \Lambda(r)$, while the fraction without reads $1 - \lambda^* = 1 - \Lambda(r)$. Here the average quality of children is $q = \lambda^* \ln(1 - \mathfrak{l} - \mathfrak{h}) + (1 - \lambda^*) \ln(1 - \mathfrak{h})$. Therefore,

$$q = \begin{cases} \lambda^* \ln(1 - \mathfrak{l} - \mathfrak{h}) + (1 - \lambda^*) \ln(1 - \mathfrak{h}), & \text{if } r = \phi \text{ (rights)}; \\ \ln(1 - \mathfrak{h}), & \text{if } r = 0 \text{ (no rights)}. \end{cases}$$

A household will vote for the law only if the wife will work. It is easy to understand why. If the wife in the household does not work, then that household will not gain any income from voting for the law. At the same time, the average quality of children in society will fall. So, the household will vote no. Recall that λ is distributed uniformly on $[0, 1]$. Again, the value for λ for median household is 0.5. If the median household votes for the law, then so will all households with a $\lambda < 0.5$. These households care less about the quality of their children than the median one does, so they are more inclined for the woman to work.

The criteria underlying the median household's vote is straightforward to formalize: the median household will vote yes (no), if the household's utility in a world with women's rights exceeds or is at least as great as (is less than) than in a world without them. If women can work (a yes vote, which implies $\lambda^* = \Lambda(\phi)$), the median household's utility will be

$$w(1 + \phi\ell) + 0.5\ln(1 - \ell - \mathfrak{h})$$

$$+ \eta\{\Lambda(\phi)\ln(1 - \ell - \mathfrak{h}) + [1 - \Lambda(\phi)]\ln(1 - \mathfrak{h})\}. \tag{1.37}$$

If they can't (a no vote, which implies $\lambda^* = \Lambda(0) = 0$), it is

$$w + 0.5\ln(1 - \mathfrak{h}) + \eta\ln(1 - \mathfrak{h}). \tag{1.38}$$

Notice that the median household cares about the average quality of children in society, which is a function of the aggregate level of married female labor-force participation, $\lambda^* = \Lambda(\phi)$—these are the terms that η multiplies in above two expressions. By using (1.37) and (1.38), one can deduce that median household will vote yes if

$$w\phi\ell - \eta\Lambda(\phi)\ln[(1 - \mathfrak{h})/(1 - \ell - \mathfrak{h})] \geq 0.5\ln[(1 - \mathfrak{h})/(1 - \ell - \mathfrak{h})],$$

and not otherwise. By solving out for $\lambda^* = \Lambda(\phi)$ using (1.36), one can rewrite the decision to vote yes ($r = \phi$) or no ($r = 0$) for married women to work as

$$r = \begin{cases} \phi \ \text{(Yes)}, & \text{if } w\phi\ell \geq 0.5\ln[(1 - \mathfrak{h})/(1 - \ell - \mathfrak{h})] + \eta w\phi\ell; \\ 0 \ \text{(No)}, & \text{if } w\phi\ell < 0.5\ln[(1 - \mathfrak{h})/(1 - \ell - \mathfrak{h})] + \eta w\phi\ell. \end{cases}$$

The left-hand side of the above expression is the gain to the median voter if the woman in the household works. She will bring home $w\phi\ell$ in additional earnings. The left-hand side is increasing trivially in ϕw. The right-hand side reflects the loss in the median's household's utility when women are allowed to work. This has two components. The first component, $0.5\ln[(1 - \mathfrak{h})/(1 - \ell - \mathfrak{h})]$, derives from the fact that the wife will spend less time with her own children if she works in the market, which results in a drop in her children's quality—note that this component can be rewritten as $0.5\ln(1 - \mathfrak{h}) - 0.5\ln(1 - \ell - \mathfrak{h})$. The second component is the loss in utility arising from the reduction in average quality of children, which turns out to be $\eta w\phi\ell$. The right-hand side is increasing in both $\mathfrak{h}$ and η.

Proposition 1 (The Evolution of Women's Rights in the Workplace)
Women's rights are more likely in societies where the requirement for housework, ħ, is low and the value of a woman's work in the market place, φw, is high. Women are likely to have fewer rights in societies that place more societal emphasis on children or equivalently where η is high.

Therefore, in a nutshell:

(1) Technological progress in the home, or a reduction in ħ, is conducive to the development of women's rights in the workplace. A model of technological progress in the home is presented in chapter 1. Note that with technological progress in the household sector, it is possible for women to go to work without spending less time with their children; this would happen when the drop in ħ is larger than ɩ. The development of women's rights in the workplace promotes labor-force participation by married females.

(2) Technological progress in the market also promotes women's rights. Technological progress in the market may arise either from an increase in the general level of wages, w, or from a decrease in the gender wage gap, as reflected by an upshift in ϕ. This could be due to a shift in the nature of jobs from blue- to white-collar (see chapter 1). Again, as a result, married female labor-force participation will rise.

2   Mothers and Sons

A boy's best friend is his mother.
—Joseph Stefano, *Psycho*

Attitudes toward married women working have changed dramatically over time. In 1938 a Gallup poll question asked, "Do you approve or disapprove of a married woman earning money in business or industry if she has a husband capable of supporting her?" Eighty one percent of men responded negatively. The General Social Survey repeated the same question in 1972, 1983, and 1998. The fraction of disapproving white males declined successively over the three surveys, from 81 percent to 38 percent, 38 to 25 percent, and 25 to 17 percent, respectively. Notions about married women working may be transmitted intertemporally through families. Economic research (based on data from the 1980s and 1990s) suggests that the probability of a man's wife working increases by somewhere between

24 to 32 percentage points if his own mother worked. This is a very large effect.

A model is now developed of how society's attitudes toward working mothers shift over time. Here, the switch in attitudes is modeled as an endogenous evolution of preferences. This transformation in preferences can be thought of as cultural change and will arise due to technological advance. There will also be a feedback effect from culture to the economy.

2.1   The Framework

The vehicle for the analysis is the model of married female labor supply along the extensive margin, developed in section 3.2 of chapter 1. Recall that married households had preferences of the form

$$u = \theta \ln c + (1 - \theta)\lambda \ln(1 - h_f) + (1 - \theta)\ln(1 - \overline{h}).$$

The husband always works the fixed amount $\overline{h}$ in the market at the wage rate w_m. The wife can either stay at home or work the fixed amount $\overline{h}$ at the wage rate w_f. Thus, the wife's labor effort, h_f, must lie in the two-point set $\{0, \overline{h}\}$. The variable λ denotes the value that a couple places on the wife's time spent at home. This differs across individuals, reflecting the fact that some families have children and others don't, or the fact that views about a woman's role at home vary across families.

Let each household live one period. Assume that they have one son and one daughter. These children will make up the households for the next period and so on. Now, suppose that the views about the value of a married woman's time spent at home differ depending on whether the husband's mother worked or not. In particular, assume that a household where the husband's mother did not work is more likely to have a high value of λ relative to one where the husband's mother did work. Operationalize this as follows: let a household where the husband's mother did not work draw $\lambda \in [\psi, \infty)$ from a Pareto distribution with cumulative distribution function

$$1 - \left(\frac{\psi}{\lambda}\right)^{\gamma}, \text{ with } \gamma > 1.$$

(The Pareto distribution is discussed in the Mathematical Appendix.) Likewise, a household where the husband's mother did work takes $\lambda \in [\xi, \infty)$ from the distribution

$$1 - \left(\frac{\xi}{\lambda}\right)^{\gamma},$$

where $\xi < \psi$. Therefore, the odds of a household drawing $\tilde{\lambda} \le \lambda$ are higher when the husband's mother worked than when she did not. This occurs because when the mother worked, $\Pr[\tilde{\lambda} \le \lambda] = 1 - (\xi/\lambda)^{\gamma}$, while when she did not, $\Pr[\tilde{\lambda} \le \lambda] = 1 - (\psi/\lambda)^{\gamma}$, and $1 - (\xi/\lambda)^{\gamma} > 1 - (\psi/\lambda)^{\gamma}$.

2.2 Married Female Labor-Force Participation

Recall from the analysis in section 3.2 of chapter 1 that there will be a threshold value for λ, denoted by λ^*, at which it makes no difference to a household whether the woman works or not. Women will not work in those households where $\lambda > \lambda^*$, because they place a high value on the women staying at home. When $\lambda \le \lambda^*$, the woman will work. By copying the earlier analysis, the household's decision regarding the wife's labor-force participation is summarized by

WORK,   if $\lambda \le \lambda^*$;

DON'T WORK,   otherwise,

where

$$\lambda^* \equiv -\frac{\theta}{1-\theta} \frac{\ln(w_m \overline{h} + w_f \overline{h}) - \ln(w_m \overline{h})}{\ln(1-\overline{h})}.$$

Thus, one can deduce that it is possible to have a situation where all women stay at home. This would occur, for example, when $w_f = 0$. When married women have always stayed at home, the lowest value of λ across household is ψ. The household with $\lambda = \psi$ is the most inclined toward the woman working. The threshold wage, w_f^*, at which it makes no difference in a household with $\lambda = \psi$ whether the wife works or not is implicitly defined by

$$\psi = -[\theta/(1-\theta)][\ln(w_m \overline{h} + w_f^* \overline{h}) - \ln(w_m \overline{h})]/\ln(1-\overline{h}).$$

(One could solve this equation for w_f^*, but there is no need to do so.) At the wage rate w_f^* no women will work, while at any higher wage rate, $w_f > w_f^*$, some married women will work.

2.3 Cultural Dynamics

Suppose that w_f jumps up permanently from the threshold wage, w_f^*, to some new higher one, $w_f' > w_f^*$, due to technological progress in the market sector. What will happen to married female labor-force participation? The new threshold value for λ will be

$$\lambda^{*'} = -[\theta/(1-\theta)][\ln(w_m \overline{h} + w_f' \overline{h}) - \ln(w_m \overline{h})]/\ln(1-\overline{h}) > \lambda^*.$$

To begin with, focus on the period when the wage first changes. At this point in time all husbands had mothers who did not work. The fraction of women working in this first period, μ_1, will be given by

$$\mu_1 = \Pr[\lambda \le \lambda^{*\prime}] = 1 - \left(\frac{\psi}{\lambda^{*\prime}}\right)^{\gamma},$$

since this gives the mass of households with a λ below $\lambda^{*\prime}$.

Next, consider the second period in time. A fraction μ_1 of husbands in the economy will have mothers who worked in the first period. These households will draw λ from the second Pareto distribution defined on $[\xi, \infty)$, as opposed to the first one defined on $[\psi, \infty)$. The fraction of married women working in the second period will be

$$\mu_2 = \underbrace{\mu_1 \left[1 - \left(\frac{\xi}{\lambda^{*\prime}}\right)^{\gamma}\right]}_{\text{MOTHERS WORKED}} + \underbrace{(1 - \mu_1)\left[1 - \left(\frac{\psi}{\lambda^{*\prime}}\right)^{\gamma}\right]}_{\text{MOTHERS DID NOT WORK}} > \mu_1,$$

where the inequality on the far right follows from the fact that $1 - (\xi/\lambda^{*\prime})^{\gamma} > 1 - (\psi/\lambda^{*\prime})^{\gamma}$. The first term in the above expression gives the proportion of married households that have working wives whose husbands had working mothers. The second term gives the fraction of households with working wives whose husbands did not have a mother who worked. Finally, moving forward in time, female labor-force participation in period t is given by

$$\mu_t = \mu_{t-1}\left[1 - \left(\frac{\xi}{\lambda^{*\prime}}\right)^{\gamma}\right] + (1 - \mu_{t-1})\left[1 - \left(\frac{\psi}{\lambda^{*\prime}}\right)^{\gamma}\right] > \mu_{t-1}.^2$$

By setting $\mu_t = \mu_{t-1} = \mu$, it is easy to see that the long-run fraction of married women working, μ, is

$$\mu = \frac{1 - (\psi/\lambda^{*\prime})^{\gamma}}{1 - [(\psi/\lambda^{*\prime})^{\gamma} - (\xi/\lambda^{*\prime})^{\gamma}]}.$$

2. This can be established by using an induction argument. Assume for the induction hypothesis that $\mu_{t-1} > \mu_{t-2}$. It then follows that $\mu_t > \mu_{t-1}$. This transpires because

$$\mu_{t-1} = \mu_{t-2}\left[1 - \left(\frac{\xi}{\lambda^{*\prime}}\right)^{\gamma}\right] + (1 - \mu_{t-2})\left[1 - \left(\frac{\psi}{\lambda^{*\prime}}\right)^{\gamma}\right].$$

After comparing the expressions for μ_t and μ_{t-1}, it follows that $\mu_t > \mu_{t-1}$ since $\mu_{t-1} > \mu_{t-2}$. To start the induction hypothesis off, note that it has been shown that $\mu_2 > \mu_1$.

Two points are worth emphasizing:

(1) The change in culture amplifies the impact of the shift in wages, which arises from technological advance, on married female labor-force participation. Without cultural change, the rise in wages would lead to an uptick in female labor-force participation from 0 to μ_1. With cultural change the long-run increase, $\mu > \mu_1$, is larger.

(2) Cultural change creates dynamics in married female labor-force participation. Married female labor-force participation rises over time: $0 < \mu_1 < \mu_2 < ... < \mu_{t-1} < \mu_t < ... < \mu$. As more mothers work, more sons have favorable attitudes toward their own wives working, and this in turn promotes more mothers working, and so on. This process may take some time and is reminiscent of Ogburn's famous cultural lag hypothesis which was advanced in his well-known 1922 book *Social Change with Respect to Culture and Natural Origin*. The cultural lag can be thought of as the time it takes between some shift in technology and the resulting adjustments in culture.

3   The Sexual Revolution, 1900–2000

There may be no better illustration of social change than the sexual revolution that occurred during the twentieth century. In 1900 almost no unmarried teenage girls engaged in premarital sex—only a paltry 6 percent (see figure 4.1). By 2002 a large majority (roughly 75 percent) were sexually experienced. What caused this was the contraception revolution. Both the technology for contraception and education about its practice changed dramatically over the course of the last century. However, notwithstanding the great improvement in contraception technology and education, the number of out-of-wedlock births to teenage girls rose from 3 percent in 1920 to 7.5 percent in 2000 (see figure 4.1). Another reflection of the change in sexual mores is the rise in the number of sexual partners that unmarried women have. For women born between 1933 and 1942, the majority of those who engaged in premarital sex had only one partner by age 20, presumably their future husband (see figure 4.2). By the 1963–1972 cohort, the majority had at least 2 partners. A little economics can go a long way toward explaining this phenomenon.

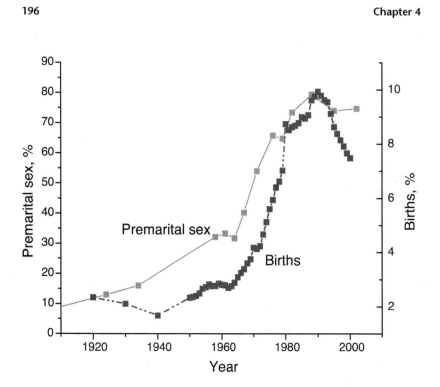

Figure 4.1
Percentage of 19-year-old women with premarital sexual experience (left axis);
teenage out-of-wedlock births, percentage (right axis).
Source: Fernandez-Villaverde, Greenwood, and Guner 2014.

4   Measures of Nonmarital Births

The following three definitions are used by demographers to measure
nonmarital (or out-of-wedlock) births. They are:

(1) *Illegitimacy Ratio*. While the term must seem antiquated now, its origin
lies in the Latin word *illegitimus*, which translates to "unlawful." The illegitimacy ratio simply gives the annual number of out-of-wedlock births per
1,000 births.

$$\text{ILLEGITIMACY RATIO} = 1,000 \times \frac{\text{out-of-wedlock births per year}}{\text{total births per year}}.$$

This is the series plotted in Figure 4.1.

(2) *Illegitimacy Rate*. This gives the fraction of unmarried women who
give birth.

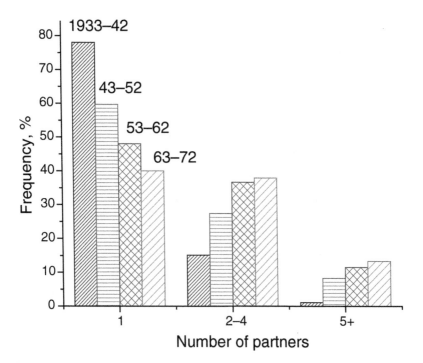

Figure 4.2
Number of partners by age 20 for women engaging in premarital sex, frequency distribution by birth cohort.
Source: Greenwood and Guner 2010.

$$\text{ILLEGITIMACY RATE} = 1{,}000 \times \frac{\text{out-of-wedlock births}}{\text{unmarried women}}.$$

Generally speaking, the illegitimacy rate is preferred to the illegitimacy ratio. For example, suppose that teenage girls almost never get married and rarely engage in premarital sex. One could find a high illegitimacy ratio for teenage girls and a low illegitimacy rate. That is, almost all births for teenage girls could be out-of-wedlock, yet there are very few of them. Sometimes it is difficult to find an estimate of the number of unmarried women, especially when doing historical work.

(3) *Prenuptial Pregnancy Ratio*. This series is often used in historical work due to a lack of data on out-of-wedlock births. By examining baptismal and marital records, historians can measure what percentage of first births

occurred within nine months following a marriage. This provides a measure of the percentage of brides who engaged in premarital sexual activity.

$$\text{Prenuptial pregnancy ratio} = 1{,}000 \times \frac{\text{first births conceived before marriage}}{\text{first births}}.$$

The use of this series to gauge illicit sexual activity is somewhat problematic, though. Marriage was a fluid concept in the past. In certain places and times, couples viewed a marriage as starting upon a formal betrothal, followed by sexual intercourse—i.e., before the nuptials. So, according to Wrigley et al. (1997, 422), "Births conceived well before marriage were probably more closely akin to illegitimate births than to those prenuptial conceptions which took place immediately before marriage. The latter class of prenuptially conceived births were arguably the product of behavior licensed by formal betrothal."

5   Technological Progress in Contraception

In 1900 engaging in premarital sex was a very risky business. Roughly 71 percent of women would have become pregnant had they engaged in sex for a year at normal frequencies using the contraception technologies at the time. These odds had dropped to 28 percent by 2002. The reduction in the chance of pregnancy occurred for two reasons: technological improvement in contraceptives and the dissemination of knowledge about contraception and reproduction. Technological improvement in contraception can result from the introduction of new contraceptives, such as the pill, or improvements in older methods, such as condoms. The spread of knowledge may encourage people to use some form of contraception in the first place, or when already using contraception to switch to more effective methods. By combining information on technological progress in contraceptives with the use of various contraceptions by women, one can construct a time series measuring the risk of pregnancy from sex. (The early data comes from birth control clinics for women, which were started at the beginning of the 1900s.)

5.1   A Brief History
Coitus interruptus has been practiced since ancient times and is mentioned in the Bible. This was the most important method of contraception historically. In addition the condom also has a long history. In the eighteenth

century, Casanova reported using the "English riding coat." Handbills cir-
culated in eighteenth-century London advertising condoms. One from Mrs.
Philips's Warehouse said:

To guard yourself from shame or fear,
Votaries to Venus, hasten here;
None in my wares e'er found a flaw,
Self-preservation's nature's law.
                    —Himes (1963, 200)

Early condoms were used more to prevent venereal diseases than pregnancies. They were expensive and uncomfortable. The diffusion of condoms
was promoted by the vulcanization of rubber in 1843–1844. They were still
expensive in 1850, selling for $5 a dozen, which translates into $34 a dozen
relative to today's real wages. So even when washed and reused, they were
too expensive for the masses. Another major innovation was the introduction of the latex condom in the 1930s, which dramatically reduced cost and
increased quality. Other methods of birth control were also used, such as a
variety of intrauterine devices. Casanova mentions using half of a lemon as
a contraceptive device. This could have been quite effective since it acted
as barrier-cum-spermicidal agent. In 1797 Jeremy Bentham advocated the
use of the sponge to keep down the size of the poor population. The rubber diaphragm entered service around 1890. It was expensive and had to
be fit by a doctor. This limited its use to those who were relatively well off.
The pill emblematizes modern contraception. It was a remarkable scientific
achievement involving the synthesis of a hormone designed to fool the
reproductive system. In 1960 the Food and Drug Administration approved
its use.

The dissemination of knowledge about contraception and reproduction
was also very important. Scientific knowledge about reproduction began to
develop in the nineteenth century. Karl Ernst Van Baer discovered the mam-
malian ovum in 1827. Around the same time, the birth control movement
in America started with the works of Robert Dale Owen and Dr. Charles
Knowlton. Owen published the first book on birth control, *Moral Physiology*,
in 1830. He suggested coitus interruptus as the best means of contraception. In 1833 Knowlton published the *Fruits of Philosophy*, which ultimately
had more influence. He advocated douching since there is "no doubt a
very small quantity of semen lodged anywhere within the vagina or within
the vulva, may cause conception, if it should escape the influence of cold,
or some chemical agent." He gave some rough prescriptions for douching

agents. Knowlton himself was prosecuted for obscenity, though. Scientific knowledge continued to progress, with George Newport describing the fertility cycle of frogs in 1853. In 1873 a law was passed, under the urging of Anthony Comstock, banning the communication, via mail, of any information about contraception or abortion. The next year the U.S. Post Office seized 60,000 rubber articles and 3,000 boxes of pills.

The modern birth control movement started around 1914, when Margaret Sanger published a pamphlet, *Family Limitation*, for which she was prosecuted. It described the use of condoms, douching, and suppositories. She became a tireless crusader for birth control clinics. She opened the first clinic in 1916. Ten days later the police came and shut it down. Apparently, a woman that the clinic had counseled was an undercover police agent. Sanger was sentenced to 30 days in prison. The first continuously effective birth control clinic was operational in 1923. Sanger promoted the use of the diaphragm through the clinics. Human ovum were seen for the first time in 1930. An accurate tracking of the ovulation cycle was also attained in the 1930s, making the safe period method a little safer. In 1952, at more than 70 years of age, Sanger persuaded a wealthy philanthropist to donate $116,000 toward the development of the pill.

Laws also evolved granting women the right to control their own fertility. Some are listed in table 4.1. In 1936, in *U.S. v. One Package*, the U.S. Court of Appeals for the Second Circuit established the right of one of Sanger's clinics to order a new type of pessary (a diaphragm) through the mail from a doctor in Japan. Dr. Hannah Stone had placed the order. Similarly, in 1965 in *Griswold v. Connecticut* the Supreme Court struck down a Connecticut law that banned people from using "any drug, medicinal article or instrument for the purpose of preventing conception." Estelle Griswold had opened a birth control clinic in November 1961. The next month the police shut it down on charges concerning the distribution of birth control. The Supreme Court ruled that Connecticut's ban on the use of contraception violated the right to marital privacy. In 1972, a women's right to an abortion was affirmed in *Roe v. Wade*. In 1969 Norma McCorvey at age 21 (aka Jane Roe) became pregnant for the third time. In an attempt to get an abortion in Texas, she lied and claimed she had been raped. She was denied an abortion due to the lack of a police report about the rape. Two crusading lawyers took the case to court on her behalf. The Supreme Court ruled that the ban on abortions violated a woman's right to privacy. Due to the time spent in litigation, Norma McCorvey actually gave birth before the decision and gave up her baby for adoption.

5.2   The Effectiveness and Use of Contraception

The use of various methods of contraception during premarital intercourse with a first partner and their efficacies are shown in tables 4.2 and 4.3. The data for contraceptive use during first premarital intercourse became available starting in the early 1960s. Between 1960 and 2002 the number of people not using any birth control fell by a remarkable 40 percentage points. The increased use of contraception may derive from two factors. First, technological improvement has made them both effective and easy to use. As more and more teenagers engage in sex because of this, one would see an increase in their use. Second, the diffusion of contraceptives may be slow, as with any new product. The birth control movement

Table 4.2

Contraception use at first premarital intercourse, percent.

Method	1900	60–64	65–69	70–74	75–79	80–82	83–88	85–89	90–94	95–98	99–02
None	61.4	61.4	54.2	55.6	53.5	46.9	34.6	36.1	29.7	27.2	21.2
Pill	-	4.2	8.6	12.1	12.8	14.2	12.1	19.7	14.1	15.3	16.0
Condom	9.42	21.9	24.0	21.0	22.0	26.7	41.8	36.4	49.9	49.8	51.2
Withdrawal	11.19	7.3	9.5	7.3	7.5	8.4	8.9	5.6	3.5	4.9	7.3
Other	17.99	5.3	3.7	4.0	4.2	3.8	2.6	2.2	2.8	2.8	4.3

Source: Greenwood and Guner 2010.

Table 4.3

Effectiveness of contraception (annual failure rates, percent).

Method	1900	60–64	65–69	70–74	75–79	80–82	83–88	85–89	90–94	95–98	99–02
None	85.0	85.0	85.0	85.0	85.0	85.0	85.0	85.0	85.0	85.0	85.0
Pill		7.5	7.5	7.5	7.5	7.5	3.4	3.4	5.5	5.5	5.5
Condom	45.0	17.5	17.5	17.5	17.5	17.5	11.0	11.0	14.5	14.5	14.5
Withdrawal	59.2	22.5	22.5	22.5	22.5	22.5	20.5	20.5	20.5	23.0	23.0
Other	50.0	20.0	20.0	20.0	20.0	20.0	20.0	10.0	10.0	10.0	10.0

Source: Greenwood and Guner 2010.

made information about contraceptives widely available (in a manner similar to advertising for other products) and access to them easy. This greatly sped up their diffusion. How much is an open question, for which it is difficult to provide a quantitative answer. The condom is the most popular method of birth control and its use has actually increased over time, notwithstanding the introduction of the birth control pill. Today more than half of people use condoms for premarital sexual relationships with their first partner. The rise of condom users played a significant role in the decline of pregnancies among the teenagers during the 1990s. The fear of catching HIV was a factor here. The upshift in the use of condoms was influenced by the expansion of formal reproductive health education during the period. Sex education about AIDS/HIV, birth control, and resisting sexual activity is associated with more consistent condom use. Furthermore, formal sex education on these topics expanded significantly during the 1990s. Chapter 5 discusses the AIDS epidemic. There a model is presented in which individuals make a decision about whether to use a condom or not when having sex.

In order to measure the decline in risk associated with premarital sex during the twentieth century, one must estimate both the use and effectiveness of contraception in 1900. Take the use of contraception in 1900, first. At this time the social sciences were in their infancy, so the analysis relies on the work of early pathbreaking researchers, such as Himes (1963) and Kopp (1934). Set the fraction of non-users in 1900 to the values observed in the 1960–1964 period (table 4.2). Clearly, this is a conservative assumption since use has been increasing steadily over time. Himes (1963) provides information on the fraction of married women who used different methods in 1930s. Assume that the selection pattern for contraception by young female users during their first premarital intercourse was similar to the pattern selected by married women. If one also assumes that the selection pattern in 1900 was the same as the one displayed in the 1930s (again a conservative assumption), contraceptive use at first premarital intercourse can be constructed for 1900.

Turn now to the effectiveness of contraception in 1900, as recorded in table 4.3. A number for effectiveness in 1900 is constructed as follows: first, Kopp (1934) reports a 45 percent failure rate for condoms and a 59.2 percent failure rate for withdrawal. His numbers are based on preclinical use by married couples who sought advice from the Birth Control Clinical

Research Bureau in New York City between 1925 and 1929. Although it seems quite high, a 45 percent failure rate for condoms is quite close to other estimates from the same period. Why was the failure rate for withdrawal so high then as well? The main reason was that partial withdrawal was considered as effective as complete withdrawal, and despite scientific evidence disproving this notion, this practice did not change quickly. Second, the other methods that people used around 1920 were not much more effective, either. Kopp (1934) reports the following failure rates: douche, 70.6 percent; jelly or suppository, 46.6; lactation, 56.6; pessary 28.1; sponge, 50; and safe period, 59.7. Hence, it is safe to presume that the failure rate for other methods at the time was no less than 50 percent. Finally, assume that using no method, and simply taking your chances, had an 85 percent failure rate, which is what the modern data show.

Since the 1960s evidence on the effectiveness of different contraceptives, for both their ideal and typical use, is quite systematic. From that time on, failure rates have been measured as the percentage of women who become pregnant during the first year of use. By contrast, the statistics from earlier studies, such as Kopp (1934), are based on married women who used birth control clinics. As can be seen, the failure rate for condoms dropped from around 17.5 percent in the 1960s to about 11 percent by the late 1980s. Somewhat mysteriously, they rose slightly in the 1990s. Overall, though, the condom's failure rate has fallen from 45 to 14.5 percent, a (continuously compounded) decline of about 113 percent, both due to technological improvement and increased knowledge about its appropriate use.

Second, as can be seen in table 4.3, the pill is the most effective method of contraception. It was introduced in the 1960s. Its initial failure rates were about 5 to 10 percent. They declined to 3.35 percent by 1989, again due to both technological improvement and better education about its use. The failure rate rose slightly during the 1990s, however. Third, even the effectiveness of withdrawal has increased over time; this shows the importance of education. Finally, during this period the effectiveness of other methods improved as well. New and much more effective methods, such as injections and implants, were introduced in the 1990s.

So what is the upshot of this analysis? By combining the information on the effectiveness and use of contraceptives contained in tables 4.2 and 4.3,

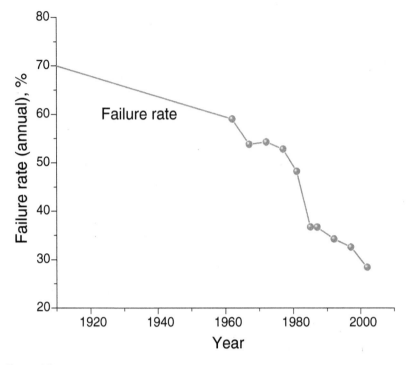

Figure 4.3
Failure rate for contraception, 1900–2002 (annualized, percent).
Source: Greenwood and Guner 2010.

one can get a measure of the extent of technological innovation in birth control. To do this, for each year take an average over the effectiveness of each method of birth control listed in table 4.3 and weight each practice by its yearly frequency of use, as shown in table 4.2. The upshot of this calculation is illustrated by figure 4.3, which presents the riskiness of premarital sex. Even when using the conservative estimates for 1900, this riskiness has fallen by (a continuously compounded) 94 percent, from about 72 percent in 1900 to 28 percent in 2002.

6   A Model of Premarital Sex

Imagine a teenage girl deciding whether or not to engage in premarital sex. When doing so, she rationally calculates the cost and benefit from this activity. The benefit is the joy from sex. Let this be related to the level of the girl's libido, denoted by l. The cost of the activity is the chance that the girl becomes pregnant. In the past, an out-of-wedlock teenage pregancy had dire economic consequences. For example in Paris during the 1860s a woman was paid about half that of a man in a similar job. Her earnings barely covered her subsistence. A working woman could earn somewhere between Fr250 and Fr600 a year, taking into account seasonal unemployment. It cost approximately Fr300 a year for rent, clothing, laundry, heat, and light. Even at the maximum salary this did not leave much for food—less than a franc a day—let alone the costs of clothing and wet-nursing a baby (the latter is estimated at Fr300 a year). A working woman in Paris at the time could certainly not afford to raise a child alone. Even today, though, a teenage pregnancy reduces the odds that a girl will attain a good education, work in a fulfilling job, or find a desirable partner on the marriage market.

Let the utility that a girl with an out-of-wedlock birth will have when she is an adult be denoted by A^o and the utility that a girl without one will realize by A^n. Assume, of course, that $A^n > A^o$. Even if the girl engages in premarital sex, she may not become pregnant. The odds of safe sex or of not becoming pregnant are represented by π. Therefore, $1 - \pi$ are the odds of becoming pregnant. The level of π reflects the state of society's contraceptive technology.

It is worth mentioning that a teenage boy might undertake the same type of cost/benefit calculation. For the boy, the cost of fathering a child out of wedlock may be much less, especially in the past before the days of paternity testing and mandated child support. The French Civil Code of 1804 prohibited questioning by the authorities about a child's paternity. (The model of premarital sex developed here is modified in chapter 5 to address the HIV/AIDS epidemic.)

6.1 A Teenage Girl's Decision-Making

Direct attention now to a teenage girl's decision about whether or not to engage in premarital sex. On the one hand, if she is abstinent, then she will realize an expected lifetime utility level of A^n. On the other hand, if she engages in premarital sex, she will realize the enjoyment l, but will become pregnant with probability $1 - \pi$. Her expected lifetime utility level will be $l + \pi A^n + (1 - \pi) A^o$. She will pick the option that generates the highest level of expected lifetime utility. Her decision can be summarized as follows:

ABSTINENCE,   if $A^n \geq l + \pi A^n + (1 - \pi) A^o$;

PREMARITAL SEX,   if $A^n < l + \pi A^n + (1 - \pi) A^0$. (6.1)

6.2   The Societal Level of Premarital Sexual Activity and Out-of-Wedlock Births

How much premarital sexual activity will there be in society? To answer this question, let libido, l, be distributed on $[0, \infty)$ according to the exponential distribution $1 - e^{-l/\beta}$, with $\beta > 0$. (A definition for the exponential distribution is provided in the Mathematical Appendix.) The exponential distribution starts at 0, when $l = 0$, and rises monotonically with l to 1, as l approaches ∞. The mean level of l is β. The threshold level of libido, l^*, at which a girl is indifferent regarding having premarital sex or not is given by

$$\underbrace{l^*}_{\text{BENEFIT}} = \underbrace{(1 - \pi)(A^n - A^o)}_{\text{EXPECTED COST}}.$$

This has a nice interpretation. The expression equates the utility of sex, given by l^*, with its expected cost, the loss in future utility induced by an out-of-wedlock birth, multiplied by the probability of pregnancy. All girls with an $l > l^*$ will engage in premarital sex while those with an $l \leq l^*$ will not. Clearly, as sex becomes safer, or when π rises, the threshold level of libido, l^*, will fall.

The fraction of sexually active teenage girls, p, is given by

$$p = 1 - \Pr[l \leq l^*] = e^{-l^*/\beta} = e^{-(1-\pi)(A^n - A^0)/\beta}.$$

Two things are worth noting about the fraction of girls engaging in premarital sex, p:

(1) This fraction moves up with an increase in π. The safer sex is, the more girls participate in it.

(2) The number of girls participating in premarital sex is decreasing in $A^n - A^o$. That is, the bigger the gap in utility between a woman without an out-of-wedlock birth and a woman with one, the less girls will engage in it. One might expect that girls from a higher socioeconomic background will have more to lose from having an out-of-wedlock birth. That is, $A^n - A^o$ will be larger for girls from a higher socioeconomic background than for ones from a lower one. In fact, in the United States, the odds of a girl having

premarital sex decline with family income. So, for instance, in the bottom decile 70 percent of girls between the ages of 15 and 19 have experienced it, versus 47 percent in the top one.

Turn now to the fraction of teenage girls who will become pregnant, o. The number of teenage pregnancies is given by the number of sexually active girls, p, multiplied by the probability that any one of them could become pregnant, $1 - \pi$. Therefore, the fraction of girls who will have an out-of-wedlock birth, o, is just

$$o = (1 - \pi)p = (1 - \pi)e^{-(1-\pi)(A^n - A^o)/\beta}.$$

This is the illegitimacy rate for teenage girls. Is o increasing or decreasing in π? To find out, take the derivative of the above expression to get

$$\frac{do}{d\pi} = \underbrace{-e^{-(1-\pi)(A^n - A^o)/\beta}}_{\text{SAFER SEX}} + \underbrace{(1 - \pi)[(A^n - A^o)/\beta]e^{-(1-\pi)(A^n - A^o)/\beta}}_{\text{INCREASE IN SEXUAL ACTIVITY}}.$$

There are two considerations here:

(1) An increase in π reduces the fraction of pregnancies, o, if the fraction of sexually active girls remains the same. This is the first term in the above expression.

(2) More girls will now experience sex, and this will raise the fraction of teenage girls who become pregnant, given the odds π for safe sex, as the second term shows.

Thus, in general, a rise in π has an ambiguous impact on the number of pregnancies. In particular, it is easy to see that

$$\frac{do}{d\pi} \gtreqless 0, \text{ as } l^* = (1 - \pi)(A^n - A^o) \gtreqless \beta.$$

Therefore, an increase in π leads to a rise (fall) in out-of-wedlock births when the threshold level of libido, l^*, lies above (below) the mean level, β.

6.3 A ∩-shaped Pattern for Teenage Pregnancies over Time

Would it be possible to see a rise and fall in out-of-wedlock births as contraception improves? The answer is yes. Start from a situation where contraception is primitive, so that π is fairly low. Historically, very few teenage girls would have been having premarital sex; i.e., p will be small.

This implies that the threshold level of libido was high, undoubtedly above the mean level. This would require $l^* = (1 - \pi)(A^n - A^o) > \beta$. Now, let π rise a little. Some more girls will be drawn into sexual activity. Since the failure rate, $1 - \pi$, is still high, a large fraction of these newly sexually active girls will become pregnant. This is the second consideration mentioned above. The first consideration has a small effect, initially. This is because the number of girls engaging in premarital sex, p, is small so that a drop in the failure rate, $1 - \pi$, does not have a large impact on pregnancies.

Let π rise further. More girls will be drawn into sexual activity. As $(1 - \pi)(A^n - A^o)$ falls, it will eventually sink below β. This must happen as π approaches 1 or as the failure rate nears 0. Then, rises in π must lead to declines in o. This occurs because when a large number of girls are sexually active (the first consideration), a drop in the failure rate leads to a fall in pregnancies. At $\pi = 1$, or when contraception is perfect, there will be no out-of-wedlock births. Therefore, to summarize, when starting out from $l^* > \beta$, out-of-wedlock births will be a $\cap$-shaped function of π, where the peak occurs at $l^* = \beta$, which implies $\pi = 1 - \beta/(A^n - A^o)$, and where the bottom of the right-hand side of the $\cap$ hits 0 at $\pi = 1$.

7   The Socialization of Children

Parents want the best for their children. They spend a great deal of time and effort educating their kids. They believe that by endowing their offspring with human capital, the latter will do better in life when they are grown up. Children must learn many things in order to succeed as adults. Not all of a person's human capital comes from formal schooling. Parents may tell their children to work hard, not to drink excessively or to do drugs, to be wise with their money, and not to be dishonest, *inter alia*. In the process, they are influencing a child's psyche. When the child grows up he or she will do whatever is in their own best interest. In economic terms, he or she will maximize his or her own utility, subject to the constraints that the person faces when an adult. But, parents can influence the path that their children will follow as adults by molding their offsprings' utility functions when they are young. Hence, the actions a child takes when an adult are partially based upon the socialization he or she received when young.

While accepted today, female premarital sex was stigmatized by families and churches in the past. So, the analysis below should be viewed within this historical context. In yesteryear having an out-of-wedlock birth had a very detrimental impact on a teenage girl's life. Families and unwed mothers could not afford to raise out-of-wedlock children. Many newborns were simply abandoned. Even today an out-of-wedlock child can reduce a teenage girl's educational and job opportunities and hurt her mating prospects on the marriage market. Parents may try to dissuade a teenage girl from entering into premarital sexual activity by inculcating in her a sense of shame associated with this activity. This was especially important in the past. Instilling a sense of worthiness from avoiding an out-of-wedlock birth—a using carrot instead of a stick, so to speak—works in exactly the same way.

Consider the problem of a lone parent who is worried about the possibility that their daughter may become pregnant when she becomes a teenager. The model to be developed builds on the analysis of the previous section with a couple of additions and one change. The daughter has a libido level of l, which the parent does not observe. Her level of libido, l, is now drawn from a uniform distribution instead of an exponential one.[3] This simplifies the analysis. As before, if she engages in premarital sex, then she will become pregnant with probability $1 - \pi$. The girl will have a utility level of A^n as an adult, if she does not have an out-of-wedlock birth. If the girl has an out-of-wedlock birth, she will realize a level of utility of A^o.

If she becomes pregnant, then the daughter will now also feel shame, in terms of utility, in the amount $-s$, where $s \geq 0$. The parent chooses the level of shame that their daughter will feel. They do this *not* knowing the libido levels, l, of the daughter. Shaming is a costly socialization process. It requires effort by the parent to instill sexual mores in their daughter. The utility cost function of this expended effort is

$$\frac{s^{1+1/\gamma}}{1+1/\gamma}, \quad \text{where } 0 < \gamma.$$

The cost of socializing the daughter is increasing in s. The marginal cost of socializing her, $s^{1/\gamma}$, also rises with s; i.e., the cost function is convex in s.

Empirically speaking, a teenage girl's feelings about the shame associated with premarital sex is related to her propensity to participate in this activity. A girl who feels that this activity is shameful is much less likely to engage in

---

3. The uniform distribution is defined in the Mathematical Appendix.

Table 4.4

Guilt and premarital sex.

Feel guilty from sex	Intercourse, %	No intercourse, %
Strongly agree	17	84
Agree	37	63
Neither agree or disagree	57	43
Disagree	75	25
Strongly disagree	77	23

Source: Fernandez-Villaverde, Greenwood, and Guner 2014 (unpublished calculations).

Table 4.5

Mother's feelings and premarital sex.

Mother's feeling about	Intercourse, %	No intercourse, %
Strongly disapprove	31	69
Disapprove	50	50
Neither approve or disapprove	81	19
Approve	76	24

Source: Fernandez-Villaverde, Greenwood, and Guner 2014 (unpublished calculations).

it, as table 4.4 makes clear. Only 17 percent of girls who would strongly feel guilt from premarital sex had sexual intercourse compared with the 77 percent who strongly said that they would not feel guilt. In a similar vein, just 31 percent of girls who felt that their mothers would strongly disapprove had sex, compared with the 81 percent who thought that their mothers would be indifferent (see table 4.5). Why would a teenage girl become sexually active if she feels guilt from it? It must be the case that the value from an intimate relationship outweighs the guilt. Peer-group pressure may also be an important factor here. This is analyzed later.

7.1 A Teenage Girl's Decision-Making, Again

On what basis does the girl decide whether or not to engage in premarital sex? Suppose a girl engages in premarital sex. She will realize the enjoyment l, but will become pregnant with probability $1 - \pi$, and she will suffer shame in the amount s. Her expected lifetime utility level when having sex will be

$l + \pi A^n + (1 - \pi)(A^o - s)$. By copying the previous analysis, one can deduce that her decision can be summarized as follows:

ABSTINENCE,   *if* $A^n \geq l + \pi A^n + (1 - \pi)(A^o - s)$;

PREMARITAL SEX,   *if* $A^n < l + \pi A^n + (1 - \pi)(A^o - s)$.

Shame reduces the expected lifetime utility from entering into a sexual relationship; i.e., it lowers the right-hand side of the above equations. It is clear that a girl will more likely practice abstinence the higher the level of shame, s, she will feel if she becomes pregnant.

Alternatively, the parent could instill a sense of fulfillment, f, if the daughter does well, which amounts here to not becoming pregnant. The girl's expected lifetime utility level when she does not have sex will now be $A^n + f$, and when she does it will appear as $l + \pi(A^n + f) + (1 - \pi)A^o$. She will choose abstinence when $A^n + f \geq l + \pi(A^n + f) + (1 - \pi)A^o$. This can be rewritten as $A^n \geq l + \pi A^n + (1 - \pi)(A^o - f)$, which is *exactly* the same as the condition involving shame. Therefore, in theory, it doesn't really matter whether a carrot, f, or stick, s, is used in the socialization process.

The threshold level of libido, l^*, for a girl reads[4]

$$\underbrace{l^*}_{\text{BENEFIT}} = \underbrace{(1 - \pi)(A^n + s - A^o)}_{\text{EXPECTED COST}}.$$

This is the libido level at which a girl is indifferent about having premarital sex or not; it sets the benefit from sexual activity equal to its expected cost. Girls with a higher level of libido will have premarital sex, and those with a lower level will not. Given that libido is uniformly distributed on $[0, 1]$, the odds that a girl will be abstinent are given by

$$\Pr[l \leq l^*] = l^*,$$

while the probability that she will have premarital sex is

$$\Pr[l \geq l^*] = 1 - l^*.$$

Therefore, the higher the level of s, the less likely it is that a girl will engage in premarital sex.

4. The threshold level of libido must lie between 0 and 1. Clearly, it will always be positive since $A^n > A^o$. Set $l^* = 1$ whenever $(1 - \pi)(A^n + s - A^o) > 1$. This is equivalent to expressing the threshold level of libido as $l^* = \min\{(1 - \pi)(A^n + s - A^o), 1\}$.

7.2 The Parent's Socialization Problem

The fraction of sexually active girls, p, is given by

$$p = 1 - l^* = 1 - (1 - \pi)(A^n + s - A^o), \tag{7.1}$$

while the proportion of girls becoming pregnant, o, can be expressed as

$$o = (1 - \pi)p = (1 - \pi)[1 - (1 - \pi)(A^n + s - A^o)].$$

This also represents the odds of a teenage pregnancy.

Suppose that the odds of a daughter becoming pregnant enter into the parent's objective function. So, too, does the cost of socialization. In particular, express the parent's decision problem as

$$\min_{s} \left\{ \rho(1 - \pi)[1 - (1 - \pi)(A^n + s - A^o)] + \frac{s^{1+1/\gamma}}{1 + 1/\gamma} \right\}.$$

The first term gives the expected loss to the parent associated with the event that their daughter becomes pregnant. The parameter ρ can be viewed as specifying the size of the cost to the parent if that occurs. The parent understands the decision process that their daughter will engage in. Specifically, they know that the probability of their daughter being sexually active and pregnant is given by $o = (1 - \pi)[1 - (1 - \pi)(A^n + s - A^o)]$. The parent also knows that this is a function of the level of shame, s, that their daughter will feel from a pregnancy. Socialization is a costly process, though. The second term in the minimization problem spells out this cost. In their decision problem, the parent weighs the expected cost of a pregnancy against the cost of socialization.

The first-order condition associated with the above problem is

$$\underbrace{\rho(1 - \pi)^2}_{\text{MB}} = \underbrace{s^{1/\gamma}}_{\text{MC}}.$$

The first-order condition sets the marginal benefit of socialization equal to its marginal cost. The marginal benefit is given by the left-hand side and shows how an extra unit of time spent socializing the girl reduces the expected cost to the parent from their daughter experiencing an out-of-wedlock birth. The right-hand side represents the marginal cost in terms of the effort that the parent spends socializing the girl. The solution for s arising from the above first-order condition is given by

$$s = [\rho(1 - \pi)^2]^\gamma. \tag{7.2}$$

Interestingly, the amount of socialization that a girl will receive is decreasing in the odds of safe sex, π. The less risky premarital sex becomes, the less need there is for a parent to socialize their daughter.

7.3   Social Change

Inserting (7.2) into (7.1) gives the fraction of girls, p, who will participate in premarital sex:

$$p = 1 - (1 - \pi)(A^n - A^o) - \underbrace{\rho^\gamma (1 - \pi)^{1+2\gamma}}_{\text{PARENTAL EFFECT}}.^5 \tag{7.3}$$

The last term in the above expression is the parental effect. This follows from a comparison of (7.1) with (7.3). The parental effect operates to reduce the number of out-of-wedlock births through shaming. The fraction of teenage girls who are sexually active, p, is unambiguously increasing in the odds of safe sex, π. As contraception becomes more effective, a larger number of girls will participate in premarital sex. It is rational for them to do so. Other things being equal, the cost of sex has fallen relative to its benefit. Additionally, parents will no longer socialize their daughters about the perils of premarital sex as much. There is less need to do so, because the odds of their daughters becoming pregnant have fallen and socialization is a costly process. This, too, is rational. A sociologist will observe a rise in premarital sex and less prohibition of it. The researcher may hypothesize that social change is causing the rise in premarital sex activity.

7.4   Religion

Other participants in the economy may also have an interest in socializing children. Churches and governments desire to curtail premarital sexual activity by teenagers. They do this for both economic and moral reasons. Governments provide welfare for unwed mothers and their children. Historically, this was a large expense for the church. Social institutions desire to minimize this expense. Teenage girls who are religious are less likely to be sexually active. A girl who says religion is very important is much less likely to have sex than one who does not think so, as is shown in table 4.6.

---

5. For the threshold level of libido to lie between 0 and 1, so must $(1 - \pi)(A^n - A^o) + \rho^\gamma (1 - \pi)^{1+2\gamma}$. Assume that this is the case.

Table 4.6
Religion and premarital sex.

Religion	Intercourse, %	No intercourse, %
Very important	38	62
Fairly important	53	47
Fairly unimportant	60	40
Not important at all	49	51

Source: Fernandez-Villaverde, Greenwood, and Guner 2014.

In yesteryear churches and civil authorities worked hard to stigmatize sex. In 1601, the Lancashire Quarter sessions in England condemned an unmarried father and mother of a child to be publicly whipped. They then had to sit in the stocks still naked from the waist upwards. A placard on their heads read "These persons are punished for fornication" (Stone 1977). In 1648, in early America, a New Haven court fined a couple for having sex out of marriage. The magistrate ordered that the couple "be brought forth to the place of correction that they may be ashamed" (Godbeer 2002). He said that premarital sex was "a sin which lays them open to shame and punishment in this court. It is that which the Holy Ghost brands with the name of folly, it is wherein men show their brutishness, therefore as a whip is for the horse and asse, so a rod is for the fool's back." These were not isolated cases. The prosecution of single men or women for "fornication" or of married couples who had a child before wedlock accounted for 53 percent of all criminal cases in Essex County, Massachusetts, between 1700 and 1785. Likewise, 69 percent of all criminal cases in New Haven between 1710 and 1750 were for premarital sex.

Illegitimacy taxed the resources of church and state. A fine, called "leyr-wite," was levied on the bondwomen of medieval English manors. The name describes its purpose and is based on the Anglo-Saxon *leger* to lie down, and *wite* a fine. This tax on fornication (6d versus a daily wage of 3/4d) levied by the lord and lady of the manor was aimed at discouraging bastardy, which placed great financial strain on the manorial community.[6] (The Church punished fornicators more ruthlessly.) A related fine was childwite, which was levied on out-of-wedlock births.

6. For more on leyrwite see Bennett (2003).

Stone (1977) relates how parish authorities in England frequently worked to ensure that bastards were born outside of their local jurisdictions, so that they would not have to absorb a financial liability. Hayden (1942–1943) discusses a similar situation in eighteenth-century Ireland, where church-wardens often employed a parish nurse, also commonly known as a "lifter." Her task was to secretly round up abandoned foundlings and deposit them in a nearby parish. Sometimes she sedated the baby with a narcotic, dia-codium, to muffle any crying. One woman, Elizabeth Hayland in the Parish of St. John's, lifted 27 babies in a year. Seven died in her care. A baby that she dropped off in the Parish of St. Paul's was promptly returned by their lifter. The churchwarden then told her not to deposit babies at the same place too often. Her salary for lifting was £3 a year. Another nurse, Joan Newenham, started out getting paid 4s 9d for every baby she lifted. This was subsequently switched to an annual salary of £4 10s.

Illegitimacy placed a great strain on the church's or state's finances. They may be called upon to provide poor relief to an unwed mother who kept her illegitimate children. They had to support the foundling hospitals and workhouses that received the abandoned babies, and provide the children with the necessary food, clothing, wetnursing, and so on. And then there was the cost of foster parents, orphanages, and workhouses for the lucky children who survived.

7.5 The Church and State

Now suppose that the church can influence the disutility parents will feel if their daughters become pregnant. Recall that ρ is the weight that a parent places on the odds of a pregnant daughter. Let

$$\rho = \frac{r^{\kappa/\gamma}}{\kappa^{1/\gamma}}, \quad 0 < \kappa < 1, \tag{7.4}$$

where r is the amount of religious indoctrination that the church under-takes. Thus, the parent will incur a higher level of disutility, the greater is the degree of religious indoctrination. Indoctrination is costly and is undertaken according to the linear cost function

$\phi r.$

Recall that the fraction of teenage girls experiencing premarital sex is given by (7.3). Multiplying this by the failure rate, $1 - \pi$, results in the proportion with an out-of-wedlock births, o:

$$o = (1 - \pi)[1 - (1 - \pi)(A^n - A^o) - \rho^\gamma (1 - \pi)^{1+2\gamma}].$$

Substituting out for ρ using (7.4) then yields

$$o = (1 - \pi) \left[1 - (1 - \pi)(A^n - A^o) - \underbrace{\frac{r^\kappa}{\kappa}(1 - \pi)^{1+2\gamma}}_{\text{Church via parents}} \right]. \tag{7.5}$$

This expression gives out-of-wedlock births, o, as a function of religious indoctrination, r. In particular, the last term in the above expression captures the church's influence on out-of-wedlock births through its impact on the amount of socialization that a parent will undertake.

The church wants to minimize the number of out-of-wedlock births, while taking into account the cost of religious indoctrination. Let the church's optimization problem be

$$\min_r \left\{ (1 - \pi) \left[1 - (1 - \pi)(A^n - A^o) - \frac{r^\kappa}{\kappa}(1 - \pi)^{1+2\gamma} \right] + \phi r \right\}.$$

There is a lot of economics encapsulated into the first term in the objective function. The church takes into account how its religious indoctrination, r, affects a parent's decision about socializing their child. It also understands how the socialization by a parent affects their teenage daughter's behavior and hence the odds of the latter having an out-of-wedlock birth. The first-order condition for religious indoctrination reads

$$\underbrace{r^{\kappa-1}(1 - \pi)^{2+2\gamma}}_{\text{MB}} = \underbrace{\phi}_{\text{MC}}.$$

The left-hand side shows how extra religious indoctrination will reduce out-of-wedlock births. The right-hand side reflects the cost of extra indoctrination. The first-order condition implies that

$$r = \left[\frac{(1 - \pi)^{2+2\gamma}}{\phi} \right]^{1/(1-\kappa)}. \tag{7.6}$$

Since $\kappa < 1$, it can be seen that as the failure rate $1 - \pi$ falls, so will the amount of religious indoctrination.

7.6   Peer-Group Effects

Teenagers are influenced by what other teenagers do. Suppose that the shame of becoming pregnant is reduced by peer-group effects, τp. That is, a girl feels less shame about a pregnancy the greater is the fraction of her peers, p, who are sexually active. How does the presence of such peer-group effects affect the above analysis?

The upshot of the teenage girl's decision problem is now

ABSTINENCE,   *if* $A^n \geq l + \pi A^n + (1 - \pi)(A^o - s + \tau p)$;

PREMARITAL SEX,   *if* $A^n < l + \pi A^n + (1 - \pi)(A^o - s + \tau p)$.

This implies that the threshold level of libido, l^*, is

$$l^* = (1 - \pi)(A^n + s - \tau p - A^o).$$

The presence of peer-group effects reduces the threshold level of libido, l^*. The fraction of girls who are sexually active, p, is now

$$p = 1 - l^* = 1 - (1 - \pi)(A^n + s - \tau p - A^o).$$

Peer-group effects operate to promote premarital sexual activity; since the shame associated with this activity is lessened, more girls engage in it.

The parent's decision problem is now

$$\min_s \left\{ p(1 - \pi)[1 - (1 - \pi)(A^n + s - \tau p - A^o)] + \frac{s^{1+1/\gamma}}{1 + 1/\gamma} \right\},$$

which has the first-order condition

$$p(1 - \pi)^2 = s^{1/\gamma}.$$

This gives the old solution for s:

$$s = [p(1 - \pi)^2]^\gamma.$$

Therefore, the parent's decision is not influenced by the presence of peer-group effects, τp. Unlike their own daughter, the parent has no control over what other girls are doing in society. That is a negative externality, just like the black smoke billowing from a factory. When making their decisions, teenage girls take the peer-group effect as given. Collectively the actions of teenagers and parents determine the peer-group effect, even though a lone teenager or parent cannot influence it.

By following the earlier analysis, one can deduce that the fraction of teenage girls that are sexually active, p, is given by

$$p = 1 - (1 - \pi)(A^n - A^o - \tau p) - \rho^\gamma (1 - \pi)^{1+2\gamma}.$$

This equation can be solved for p to get

$$p = \frac{1 - (1 - \pi)(A^n - A^o) - \rho^\gamma (1 - \pi)^{1+2\gamma}}{\underbrace{1 - (1 - \pi)\tau}_{\text{PEER EFFECT}}}. \tag{7.7}$$

The presence of peer-group effects increases the fraction of girls who are sexually active, since the denominator is decreasing in τ; the peer-group effect can be identified by juxtaposing (7.3) against (7.7).

Again, the church wants to minimize the number of out-of-wedlock births, o. As before, by engaging in religious indoctrination, r, it can affect the weight, $\rho = r^{\kappa/\gamma}/\kappa^{1/\gamma}$, that the parent places on their daughter experiencing an out-of-wedlock birth. It knows that the illegitimacy ratio for teenage girls is given by

$$o = (1 - \pi)p = (1 - \pi)[1 - (1 - \pi)(A^n - A^o) - \frac{r^\kappa}{\kappa}(1 - \pi)^{1+2\gamma}]/[1 - (1 - \pi)\tau].$$

The objective function for the church now becomes

$$\min_r \left\{ (1 - \pi) \frac{[1 - (1 - \pi)(A^n - A^o) - \frac{r^\kappa}{\kappa}(1 - \pi)^{1+2\gamma}]}{1 - (1 - \pi)\tau} + \phi r \right\}.$$

The first-order condition is

$$\frac{r^{\kappa-1}(1 - \pi)^{2+2\gamma}}{1 - (1 - \pi)\tau} = \phi,$$

which implies

$$r = \left\{ \frac{(1 - \pi)^{2+2\gamma}}{\phi[\underbrace{1 - (1 - \pi)\tau}_{\text{PEER EFFECT}}]} \right\}^{1/(1-\kappa)}. \tag{7.8}$$

The peer-group-effect term can be uncovered by contrasting (7.6) with (7.8). Therefore, the amount of religious indoctrination, r, is increasing in the strength of the peer-group effect, or in τ, at least when $\kappa < 1$. Since the presence of peer-group effects increases promiscuity in society, the church will react to this by engaging in more religious indoctrination. Unlike a lone parent, the church can affect what is happening in society at large.

The long and short of the discussion on the sexual revolution is as follows:

(1) Improvements in contraceptive practices over the course of the twentieth century can be measured, albeit imprecisely for the early years. Over time better methods of contraception have become available, and some older methods have improved. Additionally, a greater percentage of sexually active teenagers use contraception, as well as switching toward the most effective methods. This has led to a dramatic drop in the risk of pregnancy.

(2) Unmarried teenage girls rationally weigh the benefit against the expected cost of becoming sexually active. The expected cost of having premarital sex is the risk of having an out-of-wedlock birth. As this risk declines over time due to technological progress in contraception, a higher fraction of teenagers become sexually active.

(3) The initial improvements in contraceptive technology can lead to an increase in out-of-wedlock births, $(1 - \pi)p$. This occurs because these technological advances may induce a large enough hike in sexually active teenagers, p, to overwhelm the decline in the risk of pregnancy, or $1 - \pi$. As the risk of pregnancy drops to zero, so will the number of out-of-wedlock births. Thus, over time the illegitimacy ratio can display a $\cap$ shape.

(4) Parents can influence a daughter's behavior by instilling a sense of shame in her, if she has an out-of-wedlock birth. They do this knowing that the girl will act in her own best interest as a teenager. Socializing a child is a costly activity. The need to inculcate sexual mores declines with technological progress in contraception. Hence, while shaming was a major instrument for controlling a young teenager's sexual behavior in the past, it is much less so today, although vestiges of this practice still remain. The upshot is that technology affects culture. And, culture in turn influences individuals' behaviors.

(5) Churches and states can affect the degree of socialization that parents undertake with their daughters. They recognize that both daughters and parents will do what is best for themselves. The extent of religious indoctrination against illegitimacy will fall as contraception becomes more

effective. Once again, technology affects culture, with culture in turn having an impact on people's behavior.

(6) Peer-group effects operate to increase the fraction of girls that are sexually active. For parents such peer groups are a negative externality over which they have no control. Churches can affect them through religious persuasion. There will be more religious indoctrination, and hence parental socialization, in the presence of peer-group effects, other things being equal.

8 The Frequency of Sex, a Digression

The last one hundred years saw an increase in the level of premarital sexual activity. It also witnessed a rise in the frequency of sex. The earliest source on the frequency of sex is Kinsey et al. (1953). They report a mean frequency of sex for "active" women between the ages of 16 and 19 of 0.5 times per week, or 7.92 times per quarter (Kinsey et al. 1953, table 76). This classic study is based on female histories collected over the 1938 to 1950 period. Since the sample consists of women/girls between the ages of 2 and 71+, the data on premarital sexual experience provides information for the earlier periods as well. So, suppose that the frequency of sex for teenagers with premarital sexual experience was 7.92 times per quarter in 1900. Now, move forward to recent times. Data presented in Abma et al. (2004, table 6) suggests that a sexually active 15- to 19-year-old teenage girl would have had sex about 12.7 times per quarter in 2002. Therefore, the frequency of sex between participating partners rose by a factor of 1.6 over the last century.

8.1 A Model

To model the above facts, suppose that a sexually active teenage girl has the utility function of

$$\chi f^{\iota}/\iota, \text{ with } \iota < 0 \quad \text{and} \quad \chi > 0, \tag{8.1}$$

where f represents the frequency of sex. This utility function is increasing and concave in the frequency of sex, f. Let the cost of sex be given by

$$c = 1 - \pi^{f}, \tag{8.2}$$

where π is the odds of having a safe single sexual encounter. If the girl has f sexual encounters, then her odds of not becoming pregnant are given by π^f. Therefore, $1 - \pi^f$ is the probability of becoming pregnant, or the failure rate, given the frequency of sex, f. The cost function is increasing and *concave* in f, since

$$\frac{dc}{df} = - (\ln \pi) \pi^f > 0 \quad \text{and} \quad \frac{d^2c}{(df)^2} = - (\ln \pi)^2 \pi^f < 0,$$

where the signs of the above expressions follow from the fact that $0 < \pi < 1$ so that $\ln \pi < 0$. Therefore, while the chances of getting pregnant increase with the frequency of sex, they do so at a diminishing rate. Economists generally desire convex cost functions.

Cast a teenage girl's decision regarding the frequency of sex as follows:

$$\max_{f} \{\chi f^\iota / \iota - 1 + \pi^f\}.$$

The teenager seeks to maximize the benefit from sex minus its cost. The first-order condition for this problem is

$$\underbrace{\chi f^{\iota-1}}_{\text{MB}} = \underbrace{-(\ln \pi) \pi^f}_{\text{MC}}. \tag{8.3}$$

The first-order condition simply sets the marginal benefit from sex, $\chi f^{\iota-1}$, equal to its marginal cost, $-(\ln \pi)\pi^f$. The second-order condition associated with the above maximization must also be checked, because the objective function is not necessarily concave. This issue arises because the cost function is not convex. The required second-order condition for a maximum is

$$(\iota - 1)\chi f^{\iota-2} + (\ln \pi)^2 \pi^f < 0. \tag{8.4}$$

(See the Mathematical Appendix for more on maximization.)

One might expect that the frequency of sex will rise with an improvement in the effectiveness of contraceptives, or an increase in π. Strictly speaking, this need not be the case since the marginal cost of sex is decreasing in π.

Lemma 2 (The frequency of sex) *The frequency of sex, f, increases or decreases with the effectiveness of contraception, π, depending on whether $- \ln \pi \gtrless 1/f$.*

Proof   Differentiating the efficiency condition (8.3) yields

$$\frac{df}{d\pi} = \frac{-\pi^{f-1} - (\ln \pi)f\pi^{f-1}}{(\iota - 1)\chi f^{\iota-2} + (\ln \pi)^2 \pi^f}.$$

The second-order condition (8.4) implies that the denominator of the above expression is negative. Next observe that $-\pi^{f-1} - (\ln \pi)f\pi^{f-1} \lessgtr 0$, as $-\ln \pi \lessgtr 1/f$. Therefore, $df/d\pi \gtrless 0$ as $-\ln \pi \lessgtr 1/f$.   ∎

Now, for empirically relevant values of π and f, it will transpire that $df/d\pi > 0$, as the next section discusses.

8.2   Matching the Model with the Data

Is the above framework consistent with the observed increase in the frequency of sex? The answer is yes and follows from an application of the calibration methodology discussed in chapter 1. The question really amounts to asking whether or not there exist values for ι and χ such that the efficiency condition (8.3) returns the observed frequencies of sex in 1900 and 2002, given the observed probabilities of safe sex in these years. To this end, note from (8.2) that the probability of a safe single sexual encounter is given by $\pi_t = (1 - c_t)^{1/f_t}$, where the subscript t refers to the time period for a variable. The annual failure rate for contraception in 1900 was 72 percent, which converts to a quarterly rate of approximately 27 percent $= (1 - (1.0 - 0.72)^{1/4}) \times 100$ percent. The observed frequency of sex is 7.92. Hence, the probability of a safe sexual encounter in 1900 is given by $\pi_{1900} = (1 - 0.2729)^{1/7.92} = 0.9606$. Likewise, in 2002 the annual odds of not becoming pregnant were 28 percent, which converts into a quarterly probability of 5.4 percent. This implies that the odds of a safe sexual encounter in 2002 were $\pi_{2002} = (1 - 0.0543)^{1/12.71} = 0.9956$. Interestingly, while a single sexual encounter in 2002 looks very safe, having sex $4 \times 12.71 = 51$ times over the course of the year results in a 28.5 percent chance of pregnancy.

Next, it follows from (8.3) that

$$\left(\frac{f_{2002}}{f_{1900}}\right)^{\iota-1} = \frac{(\ln \pi_{2002})(\pi_{2002})^{f_{2002}}}{(\ln \pi_{1900})(\pi_{1900})^{f_{1900}}}.$$

The above equation can be used to pin down a value for ι, given observations for f_{1900}, f_{2002}, π_{1900}, and π_{2002}. Specifically, taking logs of both sides of the equation results in the following formula for ι:

$$\iota = \ln \left[\frac{(\ln \pi_{2002})(\pi_{2002})^{f_{2002}}}{(\ln \pi_{1900})(\pi_{1900})^{f_{1900}}} \right] / \ln \left(\frac{f_{2002}}{f_{1900}} \right) + 1 = -3.13.$$

Finally, a value for χ can also be backed out from (8.3). In particular, set

$$\chi = \frac{(-\ln \pi_{1900})(\pi_{1900})^{f_{1900}}}{(f_{1900})^{\iota - 1}} = 149.39.$$

To summarize, given $\iota = -3.13$ and $\chi = 149.39$, the above procedure implies that the first-order condition (8.3) will return $f = 7.92$ when $\pi = 0.9606$, and $f = 12.71$ when $\pi = 0.9956$. The second-order condition (8.4) also holds when evaluated at the 1900 and 2002 values. Thus, a maximum is obtained notwithstanding the concave cost function. With regard to Lemma 2, observe that once the framework has been calibrated to match the observed (fairly small) values for f, it must transpire that $df/d\pi > 0$, since $\ln \pi \simeq 0$.

9 The Spirit of Capitalism

A penny saved is a penny earned.
An investment in knowledge always pays the best interest.
—Benjamin Franklin

While ushering in new technologies, capital accumulation, and economic growth, the Industrial Revolution may have been spurred along by a shift in attitudes toward the virtues of patience and hard work. These values are often associated with the rise of capitalism, which in turn Max Weber connected to the earlier Protestant Reformation in his 1905 classic book *The Protestant Ethic and the Spirit of Capitalism*. A model is presented here where a change in the economic environment, caused by industrialization, leads to a transformation of values. So, unlike Weber, who connects this shift in values to an *exogenous* shift in religion, here the shift in culture is an *endogenous* response to technological progress. In particular, as the economy shifts from an agrarian one to an industrial one, there is a larger incentive to invest in human and physical capital. Investing in either human or physical capital requires a sacrifice in terms of either consumption or leisure today in order to gain greater consumption or leisure tomorrow. This brings to mind the concept of patience or time preference. The more patient people are, the more likely they are to invest in human or physical capital. As noted in the section 7, parents can try to mold their offsprings' tastes; here, they do so to favor patience. There is a cost of doing this, as parents recognize. Young adults will make their own decisions about acquiring human capital or savings, as parents know. But they also understand that by instilling patience they can change their offsprings' cost/benefit calculations. With industrialization there is a greater incentive for parents to inculcate the value of patience in their offsprings because the payoffs for their children are now greater. This is another example of how the economic environment can affect society's ideals and values—here, its attitudes toward the accumulation of wealth in terms of both financial and human capital, or capitalism more generally.

9.1   The Stanford Marshmallow Test
Patience is bitter, but its fruit is sweet.

—Jean-Jacques Rousseau

In 1960 psychologist Walter Mischel initiated a famous experiment on delayed gratification.[7] A child was offered a choice between one small reward provided immediately or double the small reward if he or she waited alone for around 15 minutes. The rewards were sometimes marshmallows (or other treats such as cookies or pretzels). Follow-up studies were conducted by Mischel and other researchers. They found that the children who managed to wait for the larger rewards tended to have better outcomes later in life, as measured by SAT scores, educational attainment, body mass index, and other life measures. For an economist, this immediately brings up the notion of time preference or discounting the future.

9.2   A Model of Instilling Self-Control
Consider a world where there are two technologies, an agricultural one and industrial one. Agricultural work demands limited or easily acquired skill. If a person works in the agricultural sector, they will earn the unskilled wage u. Industrial work requires additional skill, say either through some form of apprenticeship or education. For example, Crompton's spinning mule, which epitomizes the First Industrial Revolution, took 3 months to learn how to operate and 3 to 7 years to learn how to maintain. A job in the

---

7. A report on some of this research is Mischel, Shoda, and Rodriguez (1989).

industrial sector pays *wae*. Here *w* is the wage rate per efficiency unit of skilled labor and *ae* is the number of efficiency units that the individual supplies. The number of efficiency units that a person supplies, *ae*, is the product of their ability, *a*, and level of education, *e*. Ability and education differ across people. Let *a* be distributed across the population according an exponential distribution. Assume that a child inherits the ability level of their sole parent. A person chooses to become educated or not. Education involves a tradeoff: sacrifice today for a gain in income tomorrow. A person is more likely to invest in education the higher industrial (skilled) wages, *w*, are relative to agricultural (unskilled) ones, *u*. Additionally, the payoff from investing in education is higher the more able a person is, as reflected by *a*.

A person has two periods of life: young and old. Let a young adult's preferences be

$$1 + \beta c', \text{ with } 0 < \beta < 1,$$

where *l* is leisure when young and *c'* is consumption when old. Consumption when old, *c'*, is discounted by β. The value for the discount factor, β, is instilled in a young adult's psyche by their parent. A young adult has one unit of time that can be split between leisure, *l*, and education, *e*. The leisure enjoyed by a young person is

$$l = 1 - \frac{e^{1+1/\gamma}}{1+1/\gamma}, \text{ with } \gamma > 0,$$

where the second term on the right-hand side gives the time cost of attaining an educational level of *e*. First, the individual must decide whether or not to become educated. Second, if they decide to acquire an education, how much schooling should be obtained? These two considerations will now be analyzed, starting with the second one. As an old adult, a person spends all of their time working.

Suppose that a parent cares about the well-being of their child, as measured by the offspring's income, either *u* or *wae*. The parent recognizes that a young adult will pick the educational level that the latter wants. The parent knows that this will be a function of the young adult's discount factor, β. At the beginning of a young adult's life, the parent can pick a discount factor $\beta \geq b$ subject to a cost in terms of the parent's utility given by

$$\frac{\beta^{1+\kappa}}{1+\kappa} - \frac{b^{1+\kappa}}{1+\kappa}, \text{ with } \kappa > \gamma,$$

where b is the lower bound on child's discount factor. This cost function starts off at zero (when $\beta = b$) and rises in an ever more rapid manner with increases in β; i.e., the cost function is increasing and convex. Denote the parent's discount factor by β_{-1}. A parent knows the ability level, a, of their child.

9.3 The Decision about How Much Education to Acquire

A young adult who decides to acquire an education will solve the following maximization problem:

$$E(w, a, \beta) = \max_{e} \left\{ 1 - \frac{e^{1+1/\gamma}}{1 + 1/\gamma} + \beta wae \right\}.$$

The first-order condition for this problem is

$$\underbrace{\beta wa}_{\text{MB}} = \underbrace{e^{1/\gamma}}_{\text{MC}}.$$

The left-hand side is the marginal benefit, MB, from obtaining an extra unit of education. An extra unit of education yields an additional wa in wages. Since the payoff is in the future, it is discounted by β. The right-hand side is the marginal cost, MC, of acquiring an extra unit of education. In particular, the young adult will lose $e^{1/\gamma}$ in current leisure. Therefore the first-order condition reflects weighing a gain tomorrow in terms of additional income against the sacrifice today of lost leisure. Solving the first-order condition for e yields

$$e = (\beta wa)^{\gamma}. \tag{9.1}$$

The young adult will acquire more education the higher industrial wages, w, are, the more ability, a, they have, and the more patient they are as reflected by the discount factor β.

9.4 The Decision about Whether to Become Educated or Not

By substituting the solution for e into the young adult's objective function, an expression for the indirect utility function associated with being educated, $E(w, a, \beta)$, can be obtained:

$$E(w, a, \beta) = 1 - \frac{(\beta wa)^{1+\gamma}}{1 + 1/\gamma} + (\beta wa)^{1+\gamma}$$

$$= 1 + \frac{1}{1+\gamma}(\beta wa)^{1+\gamma}.$$

The indirect utility function, $U(u, \beta)$, connected with being uneducated is just

$$U(u, \beta) = 1 + \beta u.$$

The individual will obtain some schooling if

$$E(w, a, \beta) \geq U(u, \beta),$$

so that

$$
\begin{array}{lll}
\text{SOME SCHOOL,} & \text{if} & (\beta w a)^{1+\gamma}/(1+\gamma) \geq \beta u; \\
\text{NO SCHOOL,} & \text{if} & (\beta w a)^{1+\gamma}/(1+\gamma) < \beta u.
\end{array}
\tag{9.2}
$$

As can be seen, an individual is more likely to acquire schooling (i) the higher the industrial wage, w, is relative to the agricultural one, u; (ii) the higher is their level of ability, a; and (iii) the higher is their discount factor, β.

There will be a threshold level of ability, $\tilde{a}$, such that a young adult is indifferent concerning going to school or not. It follows from (9.2) that this threshold level of ability, $\tilde{a}$, is given by

$$\tilde{a} = [(1+\gamma)u\beta^{-\gamma}]^{1/(1+\gamma)}/w.$$

Thus, the decision to go to school can be recast as

$$
\begin{array}{lll}
\text{SOME SCHOOL,} & \text{if} & a \geq \tilde{a} = [(1+\gamma)u\beta^{-\gamma}]^{1/(1+\gamma)}/w; \\
\text{NO SCHOOL,} & \text{if} & a < \tilde{a}.
\end{array}
\tag{9.3}
$$

9.5 Instilling Patience in Children

What level of patience should a parent instill in their child? There are two cases to consider: one in which the child will go to school, and one in which the child won't. Suppose that the young adult will go to school. Then, when old, the young adult will earn

$$wae = \beta^{\gamma}(wa)^{1+\gamma}.$$

Therefore, the parent's decision problem is

$$\max_{\beta \geq b} \left\{ \beta_{-1}\beta^{\gamma}(wa)^{1+\gamma} - \frac{\beta^{1+\kappa}}{1+\kappa} + \frac{b^{1+\kappa}}{1+\kappa} \right\},$$

where again β_{-1} is the parent's discount factor. The first-order condition associated with this problem is

$$\underbrace{\beta_{-1}\beta^{\gamma-1}\gamma(wa)^{1+\gamma}}_{\text{MB}} \leq \underbrace{\beta^{\kappa}}_{\text{MC}}.$$

It will be assumed in what follows that the $\beta < 1$, although it would be easy to impose this as an additional constraint on the maximization problem.

The left-hand side of the above first-order condition for β gives the discounted marginal benefit from instilling more patience in the young adult. By raising the young adult's discount factor slightly, the parent can increase the offspring's educational level by $\gamma\beta^{\gamma-1}(wa)^{1+\gamma}$, which will in turn raise wages by $wa \times \gamma\beta^{\gamma-1}(wa)^{\gamma} = \beta^{\gamma-1}\gamma(wa)^{\gamma}$. Since this won't happen until next period, the parent discounts the gain by β_{-1}. The right-hand side gives the marginal cost to the parent in terms of utility from increasing the discount factor. This first-order condition can have either an interior or a corner solution. When an interior solution prevails, the above first-order condition will hold with equality. If the marginal benefit from instilling patience lies below its marginal cost, when the above first-order condition is evaluated at $\beta = \mathfrak{b}$, then the corner solution applies. (Again, corner solutions are discussed in the Mathematical Appendix.) The full solution for β is

$$\beta = \begin{cases} [\dfrac{1}{\beta_{-1}\gamma(wa)^{1+\gamma}}]^{1/(\gamma-\kappa-1)} = [\beta_{-1}\gamma(wa)^{1+\gamma}]^{1/(1+\kappa-\gamma)}, & \text{if } \beta \geq \mathfrak{b}; \\ \mathfrak{b}, & \text{otherwise.} \end{cases} \quad (9.4)$$

The parent will invest more in teaching a child patience (i) the higher is the industrial wage, w; (ii) the higher is the offspring's ability, a; and (iii) the higher is the parent's discount factor, β_{-1}.

If the young adult will not go to school, say because of low ability, their earnings will be u, which is not a function of the young adult's discount factor. Therefore, the parent will invest nothing, and the young adult's discount factor will be $\mathfrak{b}$.

9.6   The Steady State and Its Transitional Dynamics

What will people's discount factors be in the long run? How will the economy approach this long-run situation? From equation (9.4), it is apparent that a person's discount factor will be a function of their ability level, a. Denote this dependence by $\beta(a)$. In a steady state the discount factor for an adult of ability a must equal the discount factor for their offspring, so that $\beta(a) = \beta_{-1}(a)$. First, assume that an interior solution for the discount factor is operational for the ability level a in question. From equation (9.4), for a family of ability a, the steady-state discount factor is given by

$$\beta^*(a) = [\gamma(wa)^{1+\gamma}]^{1/(\kappa-\gamma)}.$$

Equation (9.4) specifies a first-order nonlinear difference equation. (See the Mathematical Appendix for a further discussion of this type of difference equation.) Taking logs of both sides yields

$$\ln \beta(a) = \frac{1}{1+\kappa-\gamma} \ln \gamma (wa)^{1+\gamma} + \frac{1}{1+\kappa-\gamma} \ln \beta_{-1}(a),$$

which can be expressed more simply as

$$\beta' = d + k\beta,$$

where $\beta' \equiv \ln \beta(a)$, $d \equiv [1/(1+\kappa-\gamma)] \ln \gamma (wa)^{1+\gamma}$, $k \equiv 1/(1+\kappa-\gamma)$, and $\beta \equiv \ln \beta_{-1}(a)$, with $0 < k < 1$. The steady-state level of β, or β^*, is given by

$$\beta^* = \frac{d}{1-k} = \frac{1}{\kappa-\gamma} \ln[\gamma (wa)^{1+\gamma}] = \ln \beta^*(a).$$

Now, note the following:

$$\beta' \gtreqless \beta, \text{ if and only if } \beta \lesseqgtr \beta^*,$$

and

$$\beta' \lesseqgtr \beta^*, \text{ if and only if } \beta \lesseqgtr \beta^*.$$

The above conditions are easy to verify using figure 4.4. Note that β will be negative, at least if $\beta < 1$. A steady state must lie on the 45° line, since this sets $\beta' = \beta$. Now, suppose $\beta < \beta^*$, so that β will lie to the left of β^* on the horizontal axis. The value for β' from solid line, which represents the difference equation, will lie above the 45° line. Therefore, $\beta' > \beta$. Furthermore, observe that $\beta' < \beta^*$, since on the graph β' will lie below β^*. The above two conditions imply that if $\beta < \beta^*$, then β will rise in a monotonic fashion to β^*, and if $\beta > \beta^*$, then β will fall in a monotonic fashion to β^*. Therefore, if the discount factor, $\beta(a)$, lies below (above) its steady-state level, $\beta^*(a)$, it will rise (fall) smoothly to its steady-state level.

Second, if the solution for β is at the corner solution for the value of a being considered, then the steady-state discount factor is given by $\beta^*(a) = \mathfrak{b}$.

9.7 Industrialization and the Spirit of Capitalism

Consider the impact of an unexpected one-time shift in w upward to w' due to an industrial revolution. What will happen to the steady-state discount factor for a person of ability a? Two things could potentially occur. First, their discount factor could rise from $\beta(a) = [\gamma (wa)^{1+\gamma}]^{1/(\kappa-\gamma)}$ to $\beta'(a) = [\gamma (w'a)^{1+\gamma}]^{1/(\kappa-\gamma)}$. Second, for a person of low ability, it could remain at $\mathfrak{b}$,

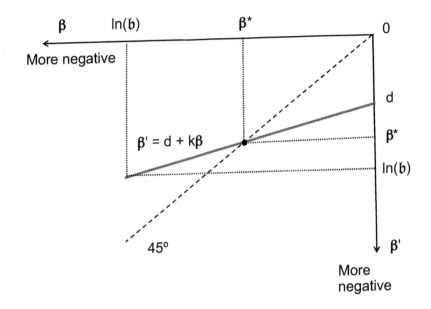

Figure 4.4
The evolution of time preference. The diagram illustrates how the first-order linear difference equation determines movement in $\beta \equiv \ln \beta_{-1}$. Note that as one moves away from the origin, either leftward on the horizontal axis or downward on the vertical one, the values become smaller or more negative. Last, since $0 < k < 1$, the slope of the line for the difference equation is less than the slope of the 45° line.

if $[\gamma (w'a)^{1+\gamma}]^{1/(\kappa - \gamma)} < \mathfrak{b}$. Only the first case needs to be discussed. Here there will be more schooling in equilibrium, both along the intensive and extensive margins. That is, there will be more people going to school since from (9.3) the threshold level of ability required to go to school will have fallen. This transpires both due to the rise in $\beta(a)$ for some people and the increase in w. Additionally, those that do go to school will acquire a higher level of education, as specified by (9.1), for similar reasons. As discussed above, the transitional dynamics will take some to work out.

10   Literature Review

The economic analysis of discrimination started with a classic book by Becker (1971). He emphasized racial discrimination. Goldin (1990) discusses marriage bars that prevented married women from working. Blau, Ferber, and Winkler (2014) have a chapter on discrimination against women in the labor force. Doepke and Tertilt (2009) provide a model of the empowerment of women over time. In their framework, in the beginning only men make decisions. The world then transits to a situation where both the men and women share power equally. Their work provided the inspiration for the model of women's rights in the workplace that is developed here. Granting married women property rights can lead to improvements in financial markets and economic growth, as is shown by Hazan, Weiss, and Zoabi (2017). Cavalcanti and Tavares (2011) find that increased labor-force participation for women is associated with larger government. In their framework public spending, such as subsidized day care, reduces the cost of having children, and this facilitates the entry of married women into the labor force. They find that empirically the size of government is positively related with female labor-force participation.

The model in section 2 is motivated by the research of Fernandez, Fogli, and Olivetti (2004), who present a model in which, following some shock to the economy that entices female labor supply, the attitudes toward working married women become more receptive over time due to the fact that more men have had working mothers. They also present evidence suggesting that a man is more likely to have a working wife if his own mother worked than if she didn't. The facts presented at the beginning of section 2 are taken from their paper. The model of premarital sex is a stripped-down version of Fernandez-Villaverde, Greenwood, and Guner (2014). The authors also did some statistical analysis on the power of shame using data from the National Longitudinal Study of Adolescent Health (Add Health). The survey asks youths questions related to shaming. The answers were aggregated into a single index using a statistical technique called *factor analysis*. The impact of shame on adolescent sexual behavior turns out to be highly significant even today. The statistical analysis controls for race, age, physical development, peer-group effects, and so on. For the language used in early America to shame premarital sex, see Godbeer (2002).

The material on the efficacy and use of contraception is taken from Greenwood and Guner (2010); the reader is directed there for further references on this subject. In the model formulated by Greenwood and Guner (2010) individuals choose a social circle that shares their beliefs about premarital sex. A partner's belief may change over time, though, as the efficacy of contraception improves. In their framework, there must be mutual consent between the partners for sex to occur. Fuchs (1992) discusses the problem of out-of-wedlock births in nineteenth-century France. Some of her facts are presented in section 6. Cavalcanti, Kocharkov, and Santos (2017) analyze the impact of lowering the cost of contraception (including abortion) in Kenya. They observe that the total fertility rate in Kenya for 2008 was 4.6. Out of this, unwanted children corresponded to 1.2 kids. They find that reducing the cost of contraception can lead to a significant increase in per capita GDP. This occurs because when parents have fewer kids, they invest more in their educations. Ermisch (2003) has a chapter on social interactions and analyzes how they might influence single parenthood.

Becker and Mulligan (1997) analyze how parents can manipulate a child's rate of time preference. Following this, Doepke and Zilibotti (2008) argue that the First Industrial Revolution triggered an socioeconomic transformation that cultivated values in the new industrialists such as frugality, thriftiness, and diligence. Section 9 is inspired by their work. For a famous paper modelling the transition from agriculture to industry, see Hansen and Prescott (2002). Finally, two survey papers on culture are Bisin and Verdier (2011) and Fernandez (2011).

11   Problems

(1) *Calibrating a model of premarital sex.* Take the framework in sections 7.1 to 7.3 and let $\gamma = 1$. In 1900 the annual failure rate for contraception was 72 percent. Only 6 percent of girls had a premarital sexual relationship at that time. By 2000 the failure rate for contraception had fallen to 28 percent. At that time, 75 percent percent of girls had premarital sex. Assume that a girl has a boyfriend for only one half of a year and is *potentially* sexually active for only one year.

(a) Is the model capable of explaining this rise in premarital sexual activity? (*Hint*: Think in terms of two linear equations in two unknown variables. Give *careful* consideration to what you would like your two variables to be.)

(b) What is the model's prediction for teenage pregnancy in the two periods? Is the assumption about how long a girl has a boyfriend important here?

(2) *Happiness and success.* A single parent has a child who will earn the wage, w, next period as an adult. The child has one unit of time that they will split when grown up between leisure, $1 - l$, and working, l. The adult child will

do this by maximizing the utility function

$$\theta \ln c + (1 - \theta) \ln(1 - l), \quad \text{with} \quad 0 < \theta < 1,$$

where c is consumption and, again, $1 - l$ is leisure. The weight that the child places on consumption is a random variable contained in a two-point set such that $\theta \in \{\underline{\theta}, \bar{\theta}\}$, with $\underline{\theta} < \bar{\theta}$. Before becoming an adult, the child will draw the high value, $\bar{\theta}$, with probability p. By socializing the child, the parent can influence this probability. The utility *cost* of socializing the child is given by

$$\sigma \left(\frac{1}{1-p} - 1 \right).$$

The parent cares about the future income that their child, as an adult, will earn. The parent discounts these future earnings at rate β.

(a) Formulate and solve the parent's socialization problem. Provide the intuition associated with your first-order condition. (Assume that the interior solution applies.)

(b) How is p connected with β, $\bar{\theta} - \underline{\theta}$, σ, and w? Explain the economics underlying your findings.

(c) Suppose instead that the parent cares about the child's happiness as an adult. The parent discounts the child's happiness at rate β. Formulate and solve the parent's socialization problem. How does the parent's attempt to influence θ change with w? What about the affect of β and σ on the socialization process? Explain the economics underlying your findings. (*Hint:* There are two cases to consider: one has only a corner solution; for the other, assume that an interior solution holds.)

5   Increased Longevity and Longer Retirement

Over the last century there was a remarkable increase in life expectancy. This was associated with a rise in the fraction of income spent on health care. A model is presented to analyze these two facts. It stresses two factors. First, technological progress in the economy lifted the overall standard of living, which raised the benefit of living longer. Second, technological advance in the medical sector lowered the cost of extending life. Developing new drugs and other medical technologies is an expensive proposition. The next section addresses this issue. It shows how rising incomes provide the wherewithal necessary to advance medicine. Additionally, it illustrates how the presence of rich patients, who finance the advancement of medicine, may help the diffusion of the new medical technologies to poorer patients. The discussion then briefly turns to the provision of health insurance and the difficulties associated with adverse selection. Loosely speaking, adverse selection occurs when the unhealthy segment of the population drives up the price of health insurance by such an extent that the healthy slice no longer wants to buy it.

The chapter then takes a detour through an important public health issue, the HIV/AIDS epidemic. According to the World Health Organization (WHO), 36.7 million people were living with the HIV virus in 2015. Around 70 percent of the infected people were in Sub-Saharan Africa. The spread of the HIV virus can be mitigated by using condoms and clean needles. The framework developed in section 6 of chapter 4 to study premarital sex is re-engineered to analyze the AIDS epidemic. In particular, the decision to engage in protected versus unprotected sex is formulated along the lines of the earlier analysis about the decision to engage in premarital sex or not. It is shown how well-intentioned public policies may actually backfire, if they do not factor in the changes in human behavior that they induce.

The last century also witnessed a dramatic rise in the fraction of life that a person will live in retirement. The next two sections analyze three forces underlying this trend. The first is the rise in the general standard of living. The second is a decline in price of the leisure goods that seniors enjoy in retirement. It is shown, by developing a simple model of retirement, how both of these factors may entice individuals to retire earlier. The third force is the provision of tax-financed old-age social security, which increases the benefit of retiring relative to working. This is analyzed by extending the baseline retirement model. The chapter concludes with a brief mention of some of the research that influenced the material presented.

1 Better Health

In 1850 a newly born American baby could expect to live just 38 years. In 1998, life expectancy was 76.7. This is a remarkable increase in life span over a relatively short time period (see figure 5.1). It is a testimonial to advances in medicine. Also important are the increases in income that make implementing the advances desirable. In particular, health-care spending's share of expenditure rises with GDP. Therefore, it appears to be a luxury good in the economics sense that it has an income elasticity greater than one. In 1929 health-care spending was 3.4 percent of GDP. This had risen to 13.1 percent by 1997. This trend is shown in figure 5.2, along with a chronology of some medical advances. Figure 5.3 shows the five-year survival rates for people diagnosed with cancer relative to the survival rate of the general population. For some cancers there has been a huge increase in the survival rate due to better detection and treatment; examples are breast, kidney, and prostrate cancers. Others such as lung cancer have seen slower progress. A byproduct of the advances in health care for humans has been a betterment in medical care for pets. Pet owners have seen their incomes rise, too. So, some of the forces at work, which explain the rise in health-care spending for humans, appear to apply for health-care spending on pets. Indeed, figure 5.4 shows that health-care spending on pets has risen in recent times as fast as health-care spending on humans. This is interesting because the institutional structures of health-care provision for humans and pets are very different.

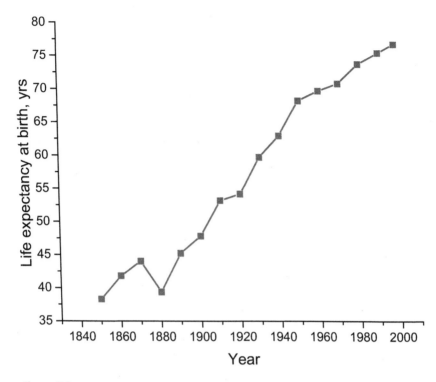

Figure 5.1

American life expectancy at birth from 1850 to 1998.

Source: Historical Statistics of the United States: Millennial Edition 2006, Series Ab644, Cambridge University Press.

1.1 A Model of Life Expectancy and Health-Care Spending

A model is now developed to analyze the rise in health-care spending and the increase in life expectancy. Two technological forces are stressed here. The first is rising wages, which are due to increases in the economy's overall level of productivity and which increase the value of living. Aggregate productivity rises because technological advance in the market sector results in a greater level of output for the marginal unit of labor hired. This makes labor more valuable, and wages rise. The second force is advances in medicine, which can be thought of as reducing the cost of extending life. An individual can potentially live for two periods. The length of each period is one unit of time. The odds of surviving into the second period are given by p. These odds are influenced by the person's spending on health care, h,

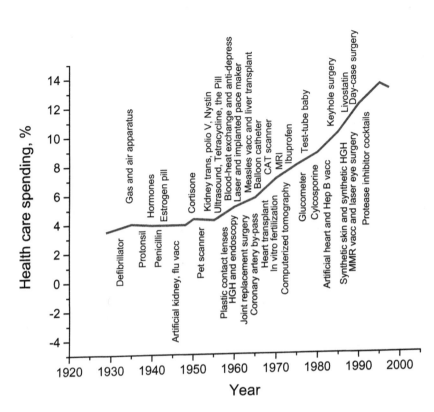

Figure 5.2
U.S. health-care spending as a fraction of GDP from 1929 to 1997, as well as a chronology of medical advances.
Sources: Historical Statistics of the United States: Millennial Edition 2006, Series Bd34, Cambridge University Press; the time line of medical advances is taken from Greenwood and Uysal 2005.

in the first period of their life. When alive, the person supplies one unit of labor and earns the wage w. This wage income can be used for consumption in the first and second periods and on health care in the first period. There is a storage technology that can transform a unit of income in the first period into $r > 1$ units of income in the second. Assume that the person's discount factor is $1/r$. An insurance market operates in the economy. Since there is an infinitely large number of people in the economy, an insurance company knows with certainty what fraction will survive and likewise what fraction will die. In particular, a unit of second-period consumption, conditional on being alive, can be purchased in the first period at the actuarially fair price

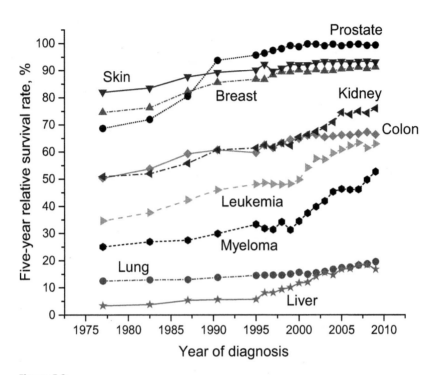

Figure 5.3
Five-year relative survival rates for people diagnosed with cancer between 1975 and 2009.
Source: SEER Cancer Statistics Review 1975–2014, Surveillance, Epidemiology, and End Results (SEER) Program, National Cancer Institute.

p/r. Likewise, a unit of second-period labor income, conditional on being alive, can be sold in the first period for p/r. That is, a certain dollar in the second period is worth $1/r$ dollars in the first period. A dollar that occurs with probability p in the second period is worth p/r dollars in first period. The size of the young population is normalized to one, which implies that there will be p old people around.

The person's expected lifetime utility function is given by

$$c + \frac{p}{r}c',$$

where c and c' are consumption in the first and second periods. Since the individual's utility is linear in c and c', the person is risk-neutral. The individual enjoys consumption in the first period with certainty. They

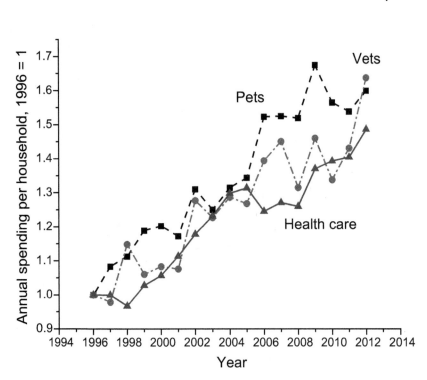

Figure 5.4
Growth in spending on health care, pet care, and veterinary care, 1996–2012. Spending in 1996 for each category has been normalized to one.
Source: Einav, Finkelstein, and Gupta 2017.

realize consumption in the second period with probability p. Second-period consumption is discounted at the rate $1/r$. With probability $1 - p$ the individual dies before the second period; the person realizes zero utility when dead.

Let the person choose the odds of surviving into the second period, p, according the health-care cost function

$$h = \frac{p^{\gamma+1}}{\eta(1+\gamma)}, \text{ where } \gamma > 0.$$

This function is increasing and convex in p. The idea is that by spending on health care in the first period, the individual can increase the odds of survival in the second period. Each increment in the odds of survival becomes evermore expensive; that is, the cost function is convex. The cost of health care drops with the parameter η, which reflects the efficiency of health-care provision. Technological progress in the health-care sector will be represented by increases in η.

The person's lifetime budget constraint appears as

$$c + p\frac{c'}{r} + h = w + p\frac{w}{r},$$

which states that the expected present value of spending on consumption and health care must equal the expected present value of labor income. The health-care cost function implies that this budget constraint can be rewritten as

$$c + p\frac{c'}{r} = w + p\frac{w}{r} - \frac{p^{\gamma+1}}{\eta(1+\gamma)}.$$

1.2 The Health-Care Decision

By substituting the budget constraint into the person's expected lifetime utility function, one can see that individual's choice problem is

$$\max_{p} \left\{ w + p\frac{w}{r} - \frac{p^{\gamma+1}}{\eta(1+\gamma)} \right\}.^{1}$$

The first-order condition for p is

$$\underbrace{\frac{w}{r}}_{\text{MB}} = \underbrace{\frac{p^{\gamma}}{\eta}}_{\text{MC}}.$$

The left-hand side gives the marginal benefit from extending life. By increasing p incrementally, the person can raise their expected earnings in the second period by w, which leads to a marginal gain in expected discounted lifetime utility of w/r. The marginal benefit of extending life is increasing in the future wage, w, that the person will earn, if alive. It is decreasing in the interest rate, r, used to discount the future. The right-hand side gives the marginal cost of raising p. This is the increase in first-period health-care spending, or p^{γ}/η. The marginal cost of extending life drops with the efficiency of health-care provision, η.

1. Strictly speaking, the probability p must satisfy the constraint $0 \leq p \leq 1$. To streamline the analysis, this constraint is ignored, and only the interior solution for the probability is entertained.

1.3 Analysis

By rearranging the above first-order condition, the odds of surviving into the second period are given by

$$p = \left(\eta \frac{w}{r}\right)^{1/\gamma}.$$

An increase in wages, w, increases the probability of living longer, p. This raises the marginal benefit from living longer because the person now earns more in the second period. Likewise, a rise in the interest rate, r, reduces p since the marginal value of working in the future falls. Technological innovation in the health-care sector, or a rise in η, also raises the probability of living longer because it reduces the marginal cost of living longer.

The person lives all of the first period with certainty. They can expect to live the fraction p of the second period. Thus, the expected life span of the person is just

$$1 + p = 1 + \left(\eta \frac{w}{r}\right)^{1/\gamma}.$$

So, anything that moves up p will increase life expectancy. The fraction of income spent on health care, f, is

$$f = \frac{h}{w + pw} = \frac{p^{\gamma+1}/[\eta(1+\gamma)]}{w + pw}.$$

The denominator of the above expression is national income, since the current young population earns w and the current old pw. The solution for p implies that $p^{\gamma+1}/\eta = pw/r$. Using this in the above equation yields

$$f = \frac{p(w/r)/(1+\gamma)}{w + pw} = \frac{1}{r(1+1/p)(1+\gamma)}.$$

This equation for f is increasing in p. It is not a function of η or w, except indirectly through p.

Proposition 1 (*Increased Longevity*) *Life expectancy, $1 + p$, and the fraction of income spent on health care, f, are increasing in wages, w, and the efficiency of health care, η.*

To conclude:

(1) A rise in wages increases both expenditure on health care and life expectancy. This occurs because there is a gain in the benefit of living longer, due to the higher standard of living in the second period.

(2) As health care becomes more efficient, as reflected by a rise in η, one should see an increase in life expectancy and a rise in the fraction of income spent on health care. This transpires because the marginal cost of investing in extending life has dropped.

Thus, the model is consistent with the facts displayed in figures 5.1 and 5.2.

2 The Development of New Drugs

Harvoni is a miracle drug that cures Hepatitis C, which is life threatening. In the United States a 12-week treatment with Harvoni costs roughly $84,000 or about $1,000 per daily pill. This is above the median household income in the United States, which was roughly $52,000 in 2014. For many people the cost is covered by insurance, which is discussed in the next section. Its price will be higher due to the cost of Harvoni. Some insurance packages do not cover (or fully cover) such expensive treatments. Some people won't buy insurance because it costs too much. Just how much it cost to develop Harvoni is unclear. Gilead Sciences, the company which sells Harvoni, spent $11 billion in 2011 to acquire Pharmasset, the pharmaceutical company that invented one of the active ingredients in the drug, sofosbuvir, which inhibits the replication of the HPC virus. Harvoni combines sofosbuvir with ledipasvir, a drug developed by Gilead Sciences, which also inhibits the replication of the virus, but in another way. After purchasing Pharmasset, Gilead Sciences spent hundreds of millions more in developing Harvoni and on the clinical trials required for FDA approval. This section models the introduction of a new drug.

2.1 The Framework

Imagine a world that goes on forever. Consumers live for one period and have a linear utility function in consumption, c, so that

$$U(c) = c.$$

A person becomes sick with probability π. Without modern medicine a sick person will die and realize zero utility. With medicine they are cured. The price of the drug, when available, is p. Individuals have different incomes. Some are poor, others are rich. This implies that some people will be able to afford the drug, while others will not. There is no market for health insurance (which is discussed later). An individual's wage is denoted by w. Let wages be distributed on $[a, b]$ according to a uniform distribution. (See the Mathematical Appendix for a definition of the uniform distribution.) The size of the population is normalized to be one.

The cost of developing the drug is δ. After it is developed, it can be produced according to the linear cost function

$$\tau q,$$

where q is the quantity of the drug manufactured and τ is its marginal cost of production. If τ falls, then so does the marginal cost of production. The developer of the drug has monopoly rights for a *single* period. Think of this as representing patent protection. After the first period, any firm can enter and produce the drug according to the above cost function; i.e., they do not have to incur the development costs. Assume that τ evolves over time according to

$$\tau' = \tau q, \tag{2.1}$$

where τ' is the next period's production cost parameter. Note that $q < 1$, because the size of the population is one, and not everyone gets sick. This implies that $\tau' < \tau$, so that the cost of producing the drug drops over time. Think about this as representing learning by doing in the production of the drug. It will be assumed that the initial marginal cost of production, τ, exceeds the poorest person's wage, a. This ensures that when the drug is first introduced, not all people will be able to afford it.

2.2 The Drug Monopolist's Problem

How should a monopolist price the new drug, assuming that it has just been developed? A person will buy the drug if $w \geq p$. In this situation, buying the drug ensures life, which yields positive utility. The size of the population that can afford the new drug is $(b - p)/(b - a)$. Only the fraction π of population that can afford the drug becomes sick, however. Therefore, the demand for the new drug, q, is given by

$$q = \pi \frac{b - p}{b - a} < 1. \tag{2.2}$$

As can be seen, demand is decreasing in price, p. The drug monopolist seeks to maximize profits. At the price p, the firm's revenue will be given by $pq = p\pi(b - p)/(b - a)$. Its production costs are $\tau q = \tau \pi (b - p)/(b - a)$.

The profit-maximizing problem facing the monopolist is

$$P = \max_{p} \left\{ p\pi \frac{b-p}{b-a} - \tau\pi \frac{b-p}{b-a} \right\} = \max_{p} \frac{\pi}{b-a} \{pb - p^2 - \tau b + \tau p\}.$$

The first-order condition for p is

$$\underbrace{2p - b}_{\text{MR}} = \underbrace{\tau}_{\text{MC}}.$$

The left-hand side is the marginal revenue, $MR = 2p - b$, gained from selling an extra unit of the drug. To see why, note that if consumers demand q units of the drug, then its price will be given by $p = b - (b-a)q/\pi$, a fact that obtains from rearranging (2.2). Therefore, the monopolist's revenue, pq, can be expressed as $pq = bq - (b-a)q^2/\pi$. This implies that marginal revenue is $b - 2(b-a)q/\pi$. By using (2.2) again, this simplifies to $2p - b$. The right-hand side is the marginal cost, $MC = \tau$, associated with producing an extra unit of the drug. Solving the above first-order condition for the price, p, gives

$$p = \frac{b+\tau}{2} > \tau = MC. \tag{2.3}$$

Observe that the monopolist sets price, p, above its marginal cost, τ, and hence is capturing rents. By contrast, in an equilibrium with perfectly competitive firms, $p = \tau = MC$. Specifically, the monopolist sets the price midway between what the wealthiest person can pay and its marginal cost. By substituting the solution for p into the monopolist's objective function, one can see that profits, P, are given by

$$P = \frac{\pi}{b-a} \left[\left(\frac{b+\tau}{2}\right) b - \left(\frac{b+\tau}{2}\right)^2 - \tau b + \tau \left(\frac{b+\tau}{2}\right) \right]$$

$$= \frac{\pi}{4(b-a)} (b-\tau)^2. \tag{2.4}$$

The lure of profits is what spurs the monopolist to develop the new drug.

2.3 The Development of the New Drug

Should the monopolist develop the new drug? Recall that the cost of developing the new drug is δ. Profits, P, must exceed the development cost, δ, for the new drug to be introduced. Or, using (2.4), it must be the case that

$$\frac{\pi}{4(b-a)} (b-\tau)^2 - \delta \geq 0.$$

As the economy gets richer due to growth, it will become possible to cover the development cost. To see this, let all wages in the economy grow over time by a constant factor. Specifically, let b and a expand by a constant factor, $\lambda > 1$. This amounts to shifting the wage distribution to the right over time. Clearly, there exists a λ such that the drug will be developed. To undertake this exercise, replace a and b by λa and λb to get

$$\frac{\pi \lambda}{4(b-a)} \left(b - \tau/\lambda\right)^2 - \delta,$$

which goes to infinity as λ increases without bound.

After the first period, imitators will enter the market, perfect competition will prevail, and the drug will be sold at marginal cost so that

$$p = \tau.$$

The price will decline over time according (2.1), because $\tau' < \tau$, as $q < 1$. So, eventually everybody will be able to purchase the new drug. Note two things. First, the drug would not be developed without the presence of patent rights. If the drug always sells at its marginal cost, τ, it is impossible to recover the development costs, δ. Second, the presence of rich patients allows the monopolist to market the new drug earlier, since these buyers provide the firm the rents necessary to fund the drug's introduction. On this last point, raise b by e and reduce a by the same amount. This holds the mean level of wages constant at $(a+b)/2$, but increases its variance from $(b-a)^2/12$ to $(b+2e-a)^2/12$. This is called a *mean-preserving spread*, which is illustrated in figure 5.5. Such a shift increases the monopolist's profits, because the right tail of the income distribution becomes richer so that a higher price can be charged for the drug; see (2.3). From (2.4) it is easy to see that the monopolist's profits (sans development costs), P, are now given by

$$P = \frac{\pi}{4(b+2e-a)} \left(b+e-\tau\right)^2.$$

With a little bit of effort, it can be shown that

$$\frac{dP}{de} = \frac{\pi}{2} \frac{(\tau - a + e)\left(b - \tau + e\right)}{(b+2e-a)^2} > 0.$$

The sign of the above expression follows from two facts. First, if the drug is being then sold, then it must be the case that $b + e > \tau$, or else not even the richest person could afford it. Second, by assumption $a - e < \tau$, which

Uniform density

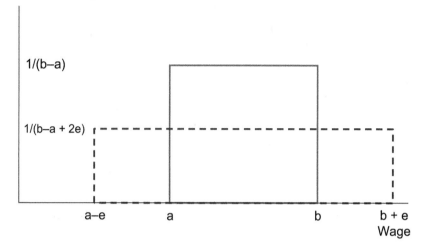

Figure 5.5
Mean-preserving spread. The picture illustrates the mean-preserving spread in the wage distribution. The rectangles show the density functions for the two uniform distributions. The distribution shifts from the initial solid rectangle to the new dashed one. The area under each rectangle is one. Both distributions have the same mean, but the dashed one has a higher variance.

guarantees that not all people will be able to afford the drug when it is first introduced.

The two takeaways from the above analysis are:

(1) Rising wages provide the resources needed to make costly advances in medicine possible. Wages rise because of productivity increases in the market sector.

(2) The lure of profits is necessary to spur innovation. The presence of rich patients who are willing to buy new medical advances at high prices stimulates innovation. Eventually, as production techniques improve and patents expire, these advances will become available to everyone.

3 Health Insurance

The issue of health insurance will now be touched upon within the context of a static model. First, it will be shown that a hale and hearty person may or may not purchase health insurance depending on its price. Second, people inclined to illness are more willing to purchase insurance at a given price than those who aren't. The presence of people on the health insurance market who are prone to illness may drive up the price of insurance to such an extent that the hale segment of the population no longer wishes to purchase it. This is called the *adverse selection problem*. When the hale and hearty segment of the population withdraws their demand for insurance, this drives up the price for the frail segment. The resulting price of insurance may then be too high even for those prone to illness to buy.

3.1 The Economy

Let a person's utility function be given by

$\ln(c + \mathfrak{c})$,

where c is their consumption and $\mathfrak{c} > 0$ is a constant. A person is healthy with probability π. A healthy person works and earns the wage w. The person can get sick with probability $1 - \pi$. When ill, the person can get treated or not. Treatment costs $\tau > w$. Hence, treatment is too expensive for a sick person to purchase. A treated person quickly recovers from their illness and works the full period. A sick person does not and dies. Observe that $\lim_{c \to 0} \ln(c + \mathfrak{c}) = \ln(\mathfrak{c})$. Thus, the individual's utility at zero consumption is finite. Take $\ln(\mathfrak{c})$ to be the value of utility at death.[2] Clearly, there is a role for health insurance in this economy. Suppose that the person can purchase health insurance at the price p. Should the individual buy health insurance?

3.2 The Decision to Purchase Health Insurance

If the person buys health insurance, expected utility is

$\pi \ln(w + \mathfrak{c} - p) + (1 - \pi) \ln(w + \mathfrak{c} - p) = \ln(w + \mathfrak{c} - p)$.

Alternatively, if they decide not to buy health insurance, expected utility reads

$\pi \ln(w + \mathfrak{c}) + (1 - \pi) \ln(\mathfrak{c})$.

2. With the log utility function, a value of zero cannot be assigned to death. For low values of $c + \mathfrak{c}$ utility will be negative. Then, a person would rather be dead than alive. Additionally, suppose that $\mathfrak{c} = 0$. Here, $\lim_{c \to 0} \ln(c) = -\infty$. For any finite value assigned to death, there would always be a low enough level of consumption such that the person would prefer death to living. The addition of the constant $\mathfrak{c} > 0$ resolves this issue.

Thus, the person's decision to buy health insurance can be summarized as

BUY,   if $\ln(w + c - p) \geq \pi \ln(w + c) + (1 - \pi) \ln(c)$;

DON'T BUY, if   otherwise.

There exists a threshold price, p^*, such that the person will buy insurance when $p \leq p^*$ and will not buy it when $p > p^*$. This price is given by

$$p^* = w + c - \exp[\pi \ln(w + c) + (1 - \pi) \ln(c)].$$

3.3   The Market Price for Health Insurance

Now, if the market for health insurance is actuarially fair, then the price of health insurance should be

$$p = (1 - \pi)\tau,$$

where the right-hand side is the just the expected payment that the health insurance company will have to make. Observe that this price reflects the cost of treatment, τ, and the probability of becoming sick. Not surprisingly, the price of health insurance rises with the cost of treatment, τ. Likewise, the price is increasing in the odds of becoming sick, $1 - \pi$. A person in an actuarially fair market will buy health insurance if

$$p^* = w + c - \exp[\pi \ln(w + c) + (1 - \pi) \ln(c)] \geq (1 - \pi)\tau.$$

In other words, the person will purchase health insurance if and only if the actuarially fair price, $(1 - \pi)\tau$, is less than or equal to the threshold price, p^*.

3.4   Adverse Selection

Suppose that there are now two types of people, namely those who get sick with probability $1 - \pi$ and those who get ill with probability $1 - \theta$, with $\theta < \pi$. The hale segment of the population are those who remain healthy with probability $\pi > \theta$. Let the fraction of the population that is hale be #. Therefore, $1 - $ # is the fraction who are frail. They have the lower odds θ of being healthy. Suppose that a person knows their type but the insurance company does not. In this situation the insurance company must charge the same price, p, to everybody. The actuarially fair price for health insurance is

$$p = [\#(1 - \pi) + (1 - \#)(1 - \theta)]\tau. \tag{3.1}$$

Note that the presence of the frail segment of population raises the price of health insurance.

The hale part of the population will buy health insurance only if

$$p \leq w + c - \exp[\pi \ln(w + c) + (1 - \pi) \ln(c)],$$

while the frail segment will only purchase it if

$$p \leq w + c - \exp[\theta \ln(w + c) + (1 - \theta) \ln(c)].$$

It is easy to see that the frail part of the population is willing to pay a higher price for health insurance, because $\theta < \pi$ and $\ln(w + c) > \ln(c)$. Now, suppose that

$$p = [\#(1 - \pi) + (1 - \#)(1 - \theta)]\tau > w + c - \exp[\pi \ln(w + c) + (1 - \pi) \ln(c)].$$

Thus, the hale slice of the population will not buy health insurance at the actuarially fair price, perhaps because the presence of the frail segment has driven the price too high. So, when insurance companies are unable to price discriminate across the hale and frail segments of the population, the actuarially fair price of insurance may be too high for the hale piece of the population to purchase it. Suppose that the hearty slice of the population does not purchase health insurance. Then, the actually fair price of health insurance is

$$p = (1 - \theta)\tau.$$

But, this price may be too high for the frail segment of the population to purchase insurance. Therefore, there will be no health insurance.

3.4.1 Single-payer mandate

Suppose the government decides to tax people the amount μ when they do not purchase health insurance. The government sets the tax μ so that a hale person is indifferent with regard to buying health insurance or not. Then, the tax is implicitly determined by the condition

$$\underbrace{\ln(w + c - p)}_{\text{Buy insurance}} = \underbrace{\pi \ln(w + c - \mu) + (1 - \pi) \ln(c)}_{\text{Don't buy insurance}}. \tag{3.2}$$

The person's utility when they buy health insurance is given by the left-hand side of (3.2). The right-hand side gives the person's expected utility when they do not buy health insurance. The presence of the tax, μ, reduces the expected utility associated with not buying the insurance. Any tax higher than the amount implicitly specified by (3.2) will result in everybody purchasing health insurance. By examining this equation, it is clear that the higher the market price of health insurance, p, is, the higher the tax, μ, will have to be. Also, the required tax is less than the health insurance premium; i.e., $\mu < p$. This can been seen by noting that $\ln(w + c - x) \geq \pi \times \ln(w + c - x) + (1 - \pi)\ln(c)$, for any value of $x < w$. Therefore, in order for the left-hand side to equal the right-hand side of (3.2), μ must be smaller than p. The tax is lower than the premium for health insurance, since purchasing health insurance yields a benefit while paying the tax does not.

4   AIDS

Two million people in the world die each year from AIDS. The problem is especially severe in Africa. The Republic of Malawi is a country where the AIDS epidemic is especially cruel. Twelve percent of the population is currently infected with the HIV virus. In Malawi HIV is transmitted primarily through heterosexual sex. (In the United States the virus is primarily passed along through homosexual sex and intravenous drug use.) The spread of AIDS in Malawi could be greatly reduced if everybody used condoms. Why don't they use condoms? Are they misinformed? No, in Malawi almost 100 percent of people are aware of HIV. Close to 60 percent of women and 75 percent of men have the knowledge that condoms lessen the chance of becoming infected. Three quarters of the population know where to obtain them. Yet, only 30 percent of women and 47 percent of men use them. People don't use condoms for two reasons. First, they believe that it reduces pleasure. One Malawian female conveyed the following sentiment about protected sex to researchers: "You can't eat (candy) while it's in the wrapper. It doesn't taste (good)." Second, within a marriage sex is often intended for reproduction, and Malawi has a much higher fertility rate than the United States. Additionally, using a condom may amount to one of the partners admitting infidelity, or the other being suspicious of it.

4.1   The Model Set-Up
The framework developed in section 6 of chapter 4 will be modified to study the AIDS epidemic. The analysis applies to the issue of sexually transmitted diseases more generally. Let people live two periods. They are young in the first period and old in the second one. In any period there are two generations alive; the current young and the current old. Think about the world as being in a stationary situation in which this period's young generation will be a facsimile, next period, of this period's old generation. Suppose

that a young healthy individual randomly meets an old partner somewhere, with whom she or he desires to have an intimate relationship. The couple has sex, which can be either protected or unprotected. Assume that old people don't care about the type of sex they have, so it will be the young person who will decide this.

With probability π the old partner has HIV, which was caught when young. This probability is an endogenous variable. It depends on the sexual behavior of people. If the old partner has HIV, then that partner will transmit it when sex is unprotected with probability μ. When sex is protected, because a condom is used, the virus is transmitted with odds ψ, where $\psi < \mu$. Whether an old person actually has HIV is *unknown*. Let H be the utility that the young person will have when older, if he or she is healthy, and A be the utility that he or she will have with AIDS.[3] Define s to be the extra utility that a young person realizes from unprotected sex. The variable $s \in [0, \infty)$ is distributed according to an exponential distribution. (The exponential distribution is defined in the Mathematical Appendix.) That is, suppose

$$\Pr[s \leq \tilde{s}] = E(\tilde{s}) = 1 - \exp(-\tilde{s}/\beta), \text{ where } \beta > 0.$$

4.2 The Decision to Have Unprotected Sex

Suppose that a young person decides to have unprotected sex. They enjoy the utility from unprotected sex, s. With probability π they have sex with someone who has HIV, which then gets transmitted with the odds μ. Hence, they will catch AIDS, and realize the utility level A, with probability $\pi\mu$. With probability $1 - \pi\mu$ they won't become sick with AIDS and will have a utility of H. Their expected utility is therefore $s + \pi\mu A + (1 - \pi\mu)H$. If the young person has protected sex, they forgo the utility s, will catch AIDS with probability $\pi\psi$, and remain healthy will odds $1 - \pi\psi$. Expected utility is then $\pi\psi A + (1 - \pi\psi)H$. In parallel with the analysis in section 6 of chapter 4, the decision about whether to have protected or unprotected sex can be represented by

PROTECTED SEX, *if* $\pi\psi A + (1 - \pi\psi)H$
$$\geq s + \pi\mu A + (1 - \pi\mu)H;$$

---

3. In 2005 antiretroviral treatment (ART) was introduced in Malawi. By 2014 about 50 percent of HIV infected individuals were on ART. To simplify things, the analysis assumes no treatment. So, in the model all HIV-infected people develop AIDS.

UNPROTECTED SEX, *if* $\pi\psi A + (1 - \pi\psi)H$

$$< s + \pi\mu A + (1 - \pi\mu)H.$$

So, the threshold level of joy from unprotected sex, s^*, is given by

$$\underbrace{s^*}_{\text{BENEFIT}} = \underbrace{\pi(\mu - \psi)(H - A)}_{\text{EXPECTED COST}} > 0.$$

This formula is intuitive. At the threshold, the joy from unprotected sex, s^*, is set equal to the expected cost of it, $\pi(\mu - \psi)(H - A)$, which is the increased loss in expected utility from catching AIDS. When $s > s^*$, the person will have unprotected sex since the joy from it is high. Alternatively, if $s \leq s^*$, the individual will have protected sex. The decision to have protected or unprotected sex depends upon the prevalence rate of the HIV virus, π, among the old population. This in turn is a function of their sexual behavior, which involves a classic externality. When engaging in unprotected sex, the current young population fails to take into account that in the aggregate they increase the HIV prevalence rate for the subsequent young generations' sexual partners.

4.3 The Presence of HIV in a Stationary Equilibrium

So, how many people will have HIV? Focus on a stationary situation in which the number of young people who become infected with HIV is the same as the fraction of old people who are infected, π. The fraction of young people having protected sex is given by

$$\Pr[s \leq s^*] = 1 - \exp(-s^*/\beta) = 1 - \exp[-\pi(\mu - \psi)(H - A)/\beta],$$

while the proportion having unprotected sex is

$$\Pr[s > s^*] = \exp[-\pi(\mu - \psi)(H - A)/\beta].$$

The people having unprotected sex will catch the disease with probability $\pi\mu$, while those having protected sex will become infected with probability $\pi\psi$. Therefore, in a stationary equilibrium, the fraction of people infected with HIV, π, will be

$$\pi = \underbrace{\pi\mu \exp[-\pi(\mu - \psi)(H - A)/\beta]}_{\text{INFECTED FROM UNPROTECTED SEX}} + \underbrace{\pi\psi\{1 - \mu \exp[-\pi(\mu - \psi)(H - A)/\beta]\}}_{\text{INFECTED FROM PROTECTED SEX}}.$$

The left-hand side of the above equation is the fraction of old people with HIV. The right-hand side is the proportion of young people who become infected. This has two sources: unprotected sex, the first term, and protected sex, the second one. Take the first source. The fraction $\exp[-\pi(\mu - \psi)(H - A)/\beta]$ of young adults have unprotected sex with old adults. An old adult has HIV with probability π. This gets transmitted to a young adult having unprotected sex with probability μ. So, $\pi\mu\exp[-\pi(\mu - \psi)(H - A)/\beta]$ is the fraction of young adults who become infected with HIV from unprotected sex. The second source can be explained in a similar manner. Observe that $\pi = 0$ is a solution to this equation. If no old people have HIV, then no young people can become infected. There is another solution, too. To see this, first divide through the above equation by π and then rewrite it as

$$(\mu - \psi)\exp[-\pi(\mu - \psi)(H - A)/\beta] = 1 - \psi.$$

This new equation defines an implicit solution for π where HIV is present in the steady state. Observe that $\pi = 0$ cannot be a solution to the above equation, because then the left-hand side would be smaller than the right-hand one.

4.4 The Effects of Public Policy

Now, consider the impact of a public policy intervention that reduces μ, or the risk of transmission with unprotected sex. Some experts believe that male circumcision does this, because foreskin is highly vulnerable to HIV infection. What will be its impact? To analyze this, take the derivative of π with respect to μ in the above expression.[4] One obtains

$$\frac{d\pi}{d\mu} = \frac{1 - (\mu - \psi)\pi(H - A)/\beta}{(\mu - \psi)^2(H - A)/\beta}.$$

so that

$$\frac{d\pi}{d\mu} \gtrless 0, \text{ as } \underbrace{\beta - (\mu - \psi)\pi(H - A)}_{s^*} \gtrless 0.$$

Therefore, an *increase* in μ can lead to a rise or fall in HIV prevalence, depending on whether $\beta \gtrless s^*$. When $\beta > s^*$, the mean level of enjoyment for unprotected sex lies above the threshold level. An increase (decrease) in μ will then lead to a rise (fall) in HIV. If $\beta < s^*$, the reverse is true.

---

4. To do this totally differentiate the above expression with respect to π and μ. The total derivative is discussed in the Mathematical Appendix.

There are two offsetting effects at work here. First, it is true that if μ becomes smaller, then HIV will decrease, other things equal. But, other things are not equal and this leads to the second effect. A drop in μ also results in more people having unprotected sex. When $\beta > s^*$, a substantial number of people are already having unprotected sex, because the threshold for having unprotected sex lies below the population mean. So, the impact of a reduction in μ, which makes sex safer, will dominate. When $\beta < s^*$, the pool of people having unprotected sex is smaller. So the impact of making sex safer on the prevalence of HIV is diminished, while the increase in the fraction of people having unprotected sex is proportionately bigger. Now, the second effect dominates, and the prevalence of HIV increases. So, it is important for epidemiologists to take behavioral reactions into account when making policy prescriptions.

5   The Trend in Retirement

Retirement is a recent phenomenon. In 1880 more than 75 percent of men over the age of 65 were still working. Contrast this with only 20 percent in 2000. This is not only an American phenomenon. The labor-force participation rates for men older than 65 have declined continuously over the last 100 years in France, Germany, and Great Britain as well. Figure 5.6, left panel, plots the retirement rates over time for men of various ages in the United States. In order to be classified as retired, a person must be completely out of the labor force. Observe the dramatic increase in retirement rates for older men. For example, in 1850 only 20 percent of men aged 75–79 were retired. By 2000 nearly 90 percent were. The right panel plots the fraction of a person's life spent in retirement. In 1850 a 20-year-old could have expected to live just 6 percent of his life in retirement, compared with roughly 30 percent in 2000. The focus here on men derives from the fact that historically the labor-force participation rates for women were much lower than for men. As can be seen from the right-hand-side panel of figure 1.13 in chapter 1, only 2 percent of married women worked in 1900, which moved up to 21 percent in 1950; the majority still did not work in 1980. To be counted as retired, statistically speaking, you must have worked in the market.

What do retired men do? The uses of time are documented in table 5.1 for various age groups of men. Men age 65 or older spend nearly 43 percent

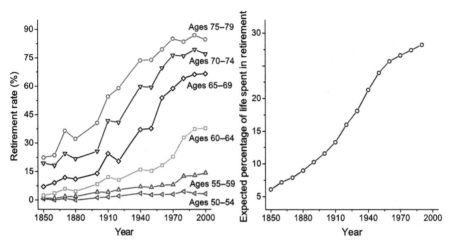

Figure 5.6
The trend in retirement. The left panel charts retirement rates for American men older than age 50 for the period 1850–2000. The right panel plots the expected percentage of adult life spent in retirement for a 20-year-old American male.
Source: Kopecky 2011, figure 2.

Table 5.1
Weekly hours spent by men on leisure activities in 1985.

	Age		
Activity	25–54	55–64	65+
Participating in organizations	0.9	2.2	1.1
Attending events	0.9	0.6	0.1
Visiting	6.6	6.7	6.0
Playing or watching sports	2.9	2.8	3.0
Hobbies	2.5	3.5	3.5
Talking or socializing	2.8	3.1	4.5
Watching TV	16.1	18.2	24.9
Reading	2.6	4.7	6.7
Listening to music	0.5	0.9	1.3

Source: Kopecky 2011, table 2.

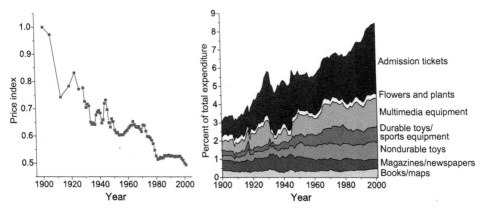

Figure 5.7
Leisure goods. The left panel shows the relative price of leisure goods while the right panel gives their share of total expenditure for the period 1900–2001 in the United States.
Source: Kopecky 2011, figure 3.

more time on leisure activities than do men aged 25–54. They use this time to listen to music, read books, and watch TV. Upon retirement, individuals allocate a larger share of their budget to leisure goods. Figure 5.7, left panel, shows how the price of leisure goods has fallen at approximately 1 percent a year over the last 100 years. Associated with this decline has been a rise in the fraction of spending allocated to leisure goods. This is shown in the right-hand-side panel of figure 5.7. In 1900 Americans spent about 3 percent of their budgets on leisure goods. This had risen to 8 percent by 2000.

Last, poorer countries today may partially mirror the United States of yesteryear. The cross-country relationship between market hours worked for people 55 or older and the natural logarithm of per-capita GDP is charted in figure 5.8. There is a strong negative relationship between the two variables; the Pearson correlation coefficient is -0.78. Older people work much less in wealthier countries. Retirement appears to be a luxury good.

The analysis below stresses retirement as a block of time that individuals use to enjoy leisure goods. As the price of such goods declined over time and as real wages grew, parceling off such a block of time to enjoy leisure activities became more affordable. This created an incentive for people to spend more time in retirement.

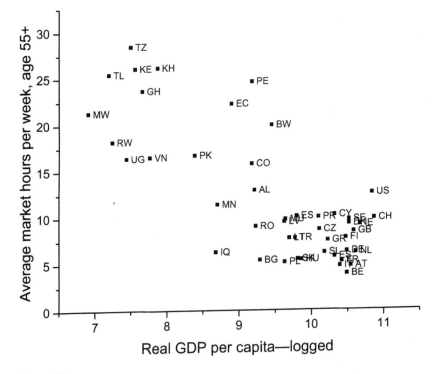

Figure 5.8
The cross-country relationship between average market hours per week for people age 55 or older and the logarithm of GDP per capita.
Source: As tabulated by Bick, Fuchs-Schundeln, and Lagakos 2018.

5.1   The Model's Scaffolding

Imagine a person whose life span is the interval [0, 1]. The individual can split this interval between working and retiring. At each point on the interval the person has one unit of time. When the person works, this one unit of time is used on the labor market and earns the wage w. The person can use their labor income to consume and save at the interest rate r (which is assumed to be greater than the individual's subjective rate of time preference). There are two types of goods: a general consumption good, c, and a leisure good, ρ. General goods yield an instantaneous utility of $\ln(c)$. If the person is retired, then their unit of time is used for leisure. In particular, this unit of time is mixed with a leisure good that generates a fixed utility level of ρ. Leisure goods are time intensive. In order to enjoy a television you must spend time watching TV shows. That is, leisure goods exhibit a high degree of complementarity between the good and the user's time. In the current setting, think about the leisure good as requiring one unit of time to enjoy. The leisure good can be purchased at the price p.

5.2 The Retirement Decision

The period-0 discounted lifetime utility for the person is given by

$$\int_0^1 \ln c(t) \exp(-\beta t)dt + \rho \int_R^1 \exp(-\beta t)dt,$$

where $0 < \beta < 1$ is the subjective rate of time preference, $c(t)$ is time-t general consumption, ρ is the fixed instantaneous utility from the leisure good, and R is the date of retirement. The individual discounts future utilities using their subjective rate of time preference, β, which is taken to be less than the market interest rate, r, so that $\beta < r$. Note that $\exp(-\beta t) < 1$, for all $t > 0$, and is decreasing in time, t. The person only consumes the leisure good during retirement or on the interval $[R, 1]$. The individual's budget constraint reads

$$\int_0^1 c(t) \exp(-rt)dt + \int_R^1 p \exp(-rt)dt = \int_0^R w \exp(-rt)dt. \tag{5.1}$$

Here the person works on the interval $[0, R]$ and again consumes the leisure good only on $[R, 1]$. The left-hand side of the budget constraint gives the present value of the spending on consumption goods, both general and leisure. The right-hand side is the present value of labor income. These present values are calculated using the market rate of interest, r, because this is the rate at which the individual can borrow or lend.

Let λ be the period-0 Lagrange multiplier associated with the person's maximization problem. (The Lagrange multiplier is reviewed in the Mathematical Appendix.) The Lagrangian maximization problem can be stated as

$$\max_{c(t),R} \left\{ \int_0^1 \ln c(t) \exp(-\beta t)dt + \rho \int_R^1 \exp(-\beta t)dt \right.$$

$$\left. + \lambda \left[\int_0^R w \exp(-rt)dt - \int_0^1 c(t) \exp(-rt)dt - p \int_R^1 \exp(-rt)dt \right] \right\}.$$

The first-order conditions for $c(t)$ and R that are associated with this maximization problem are

$$\underbrace{\frac{1}{c(t)}\exp(-\beta t)}_{\text{MB}} = \underbrace{\lambda\exp(-rt)}_{\text{MC}},$$

and

$$-\rho\exp(-\beta R) + \lambda w\exp(-rR) + \lambda p\exp(-rR) \geq 0,$$

where Leibniz's rule for differentiating an integral is used to obtain the second one. (Leibniz's rule is presented in the Mathematical Appendix.) Rewrite the second one as

$$\underbrace{\rho\exp(-\beta R)}_{\text{MB}} \leq \underbrace{\lambda(p+w)\exp(-rR)}_{\text{MC}}. \tag{5.2}$$

These two first-order conditions have natural interpretations. The first states that the marginal benefit from consuming a bit extra in period t must equal its marginal cost. The marginal benefit of an extra unit of period-t general consumption is its discounted period-t marginal utility, $\exp(-\beta t)\mathrm{MU}_{c(t)} = \exp(-\beta t)/c(t)$. Turn to the marginal cost side of the equation. The Lagrange multiplier, λ, gives the marginal utility from an extra unit of time-0 income, which is equal to the marginal utility of time-0 general consumption. To see this, set $t = 0$ in the first-order condition and note that $\mathrm{MU}_{c(0)} = 1/c(0) = \lambda$. A unit of period-$t$ income is worth $\exp(-rR)$ in terms of period-0 income. Therefore, $\lambda\exp(-rt)$ is the marginal cost of an extra unit of period-t general consumption.

The second first-order condition sets at the date of retirement, R, the discounted utility from consuming the leisure good to be less than or equal to its cost. This discounted utility from consuming the leisure good at the date of retirement is $\rho\exp(-\beta R)$. The cost is made up of two parts; $\lambda w\exp(-rR)$ and $\lambda p\exp(-rR)$. First, by retiring a split-second earlier and consuming the leisure good, the individual loses the wage, w, at the date of retirement, R. The present discounted value of this in terms of current marginal utility is $\lambda w\exp(-rR)$. Second, the person also incurs the cost of the leisure good, p. This costs $\lambda p\exp(-rR)$ in terms of current marginal utility. The person will never not work; i.e., they will never set $R = 0$ because the utility from general goods would be $-\infty$. They might never retire, though. That is, the individual might set $R = 1$, if $\rho\exp(-\beta) < \lambda(p+w)\exp(-r)$. Here, even at the last instant of life ($R = 1$), the marginal benefit of retiring a moment later lies below its marginal cost. This could happen when ρ is low. In what follows assume that an interior solution for retirement always occurs.

Now, using the fact that $c(0) = 1/\lambda$ in the first-order condition for general consumption, one can deduce that

$$c(t) = c(0) \exp[(r - \beta)t].$$

This is a well-known result. Consumption grows over time in line with the gap between the real interest rate, r, and the subjective rate of time preference, β. Note that $\exp[(r - \beta)t] > 1$, when $r > \beta$ and $t > 0$, and is increasing in t. For general consumption, $c(t)$, to rise over time, the person must be saving; hence, there is a savings decision embedded in the above optimization problem. This should be obvious because how else could the person survive in retirement? When $r > \beta$, the rate of return on savings, r, earned by delaying some consumption, exceeds the rate of return that the individual requires to postpone some consumption, as dictated by the rate of time preference, β. This favors some postponement so general consumption grows over time. Plugging this solution for $c(t)$ into the budget constraint (5.1) and integrating then yields

$$\frac{c(0)}{\beta}[1 - \exp(-\beta)] = \frac{w}{r}[1 - \exp(-Rr)] - \frac{p}{r}[\exp(-Rr) - \exp(-r)].^5 \qquad (5.3)$$

Next, substituting the fact that $c(0) = 1/\lambda$ into the second first-order condition and solving for $c(0)$ gives

$$c(0) = \frac{p + w}{\rho} \exp[-(r - \beta)R].$$

Inserting this into the previous equation, and then dividing through by w, results in

$$\frac{p/w + 1}{\beta\rho} \exp[-(r - \beta)R][1 - \exp(-\beta)] = \frac{1}{r}[1 - \exp(-Rr)]$$

$$-\frac{p/w}{r}[\exp(-Rr) - \exp(-r)]. \qquad (5.4)$$

This is a single nonlinear equation in R. It is analyzed now.

5. Note that $\int_a^b \exp(-\iota t)dt = -(1/\iota) \exp(-\iota t)|_a^b = (1/\iota)[\exp(-\iota a) - \exp(-\iota b)]$. Therefore, $\int_0^1 \exp(-\beta t)dt = (1/\beta)[1 - \exp(-\beta)]$, $\int_0^1 \exp(-rt)dt = (1/r)[1 - \exp(-r)]$, and $\int_R^1 \exp(-rt)dt = (1/r)[\exp(-Rr) - \exp(-r)]$.

5.3 Analysis of the Retirement Decision

Observe that the left-hand side of (5.4) is decreasing in R, while the right-hand side is increasing. The situation is portrayed in figure 5.9. Additionally, the left-hand side is increasing in the time price for leisure goods, p/w, while the right-hand side is decreasing in p/w. A increase in p/w to p'/w' results in the left-hand-side curve shifting up and the right-hand-side one shifting down. As a result, retirement is postponed.

Proposition 2 (*The Trend in Retirement*) *A drop in the time price of leisure goods, p/w, leads to earlier retirement or a decline in R.*

The proposition implies that people will retire earlier following:

(1) a drop in the price of leisure goods, p,

(2) a rise in the wage rate, w.

Note that a *decline* in p/w will cause the left-hand-side curve to shift *down* and the right-hand-side one to move *up*. Thus, R falls. As leisure goods become more affordable, it makes sense to increase the block of time, here

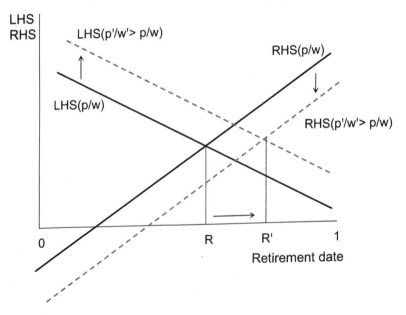

Figure 5.9
Determination of retirement. The diagram illustrates how an increase in the time price of leisure goods from p/w to p'/w' delays retirement from R to R'.

retirement, used to enjoy them. Thus, the model's predictions are in line with the trends in retirement displayed in figure 5.6.

6   Old-Age Social Security

The retirement model developed in the previous section will now be used to analyze the provision of tax-financed old-age social security. Suppose that the government decides to tax a person's labor income at the rate τ and use the proceeds to provide an old-age pension in the lump-sum amount s. It will be assumed that all individuals are identical, with the size of the population being normalized to one. How does this provision of old-age social security affect a person's decision to retire? It will be shown that this decreases the marginal cost of retirement. Hence, the Social Security Act passed in 1935 could account for some of the increase in retirement. It is important to note however, that the trend in retirement started well before 1935, as figure 5.6 illustrates. Therefore, the technological considerations discussed previously are important and are probably the dominant factors. The analysis below abstracts from certain features of the U.S. Social Security system, such as the fact that you can only collect benefits after a certain age (currently 62) and that the benefits increase with the age chosen for retirement (until age 70).

6.1   The Retirement Decision, Again
The individual's budget constraint now appears as

$$\int_0^1 c(t)\exp(-rt)dt + \int_R^1 p\exp(-rt)dt = \int_0^R (1-\tau)w\exp(-rt)dt$$
$$+ \int_R^1 s\exp(-rt)dt. \tag{6.1}$$

The person will now take home $(1-\tau)w$ in after-tax labor income for each period in the interval $[0, R]$ that they work. They will receive an old-age pension, s, in the interval $[R, 1]$ when they do not work. The individual's maximization problem is

$$\max_{c(t),R} \left\{ \int_0^1 \ln c(t)\exp(-\beta t)dt + \rho \int_R^1 \exp(-\beta t)dt + \lambda \left[\int_0^R (1-\tau)w\exp(-rt)dt \right. \right.$$
$$\left. \left. + \int_R^1 s\exp(-rt)dt - \int_0^1 c(t)\exp(-rt)dt - \int_R^1 p\exp(-rt)dt \right] \right\},$$

where again λ is the period-0 Lagrange multiplier. The two first-order conditions connected to this problem are

$$\frac{1}{c(t)}\exp(-\beta t) = \lambda\exp(-rt),$$

and

$$\underbrace{\rho\exp(-\beta R)}_{MB} = \underbrace{\lambda[p + (1-\tau)w - s]\exp(-rR)}_{MC}. \tag{6.2}$$

The one for consumption is unchanged, while the one connected with the retirement decision is changed. The marginal benefit from retiring earlier is the same as before. This is the discounted utility from consuming the leisure good at the date of retirement. The marginal cost term is now different. By retiring a moment earlier, the individual now loses only their after-tax wage, $(1-\tau)w$, at the date of retirement, R, instead of the wage, w. The person also gains an old-age pension, s, which works to offset the loss in labor income. Both of the factors reduce the marginal cost of retirement.

To complete the description of the economy, suppose that the present value of the taxes that the government collects from individuals exactly equals the present value of the old-age benefits that they will receive; i.e., the government's budget is balanced. This implies that

$$\int_0^R \tau w \exp(-rt)dt = \int_R^1 s\exp(-rt)dt. \tag{6.3}$$

Call this condition the government's budget constraint. By performing the integration on both sides, this equation implies that the old-age pension, s, can be written as

$$s = \tau w\left[\frac{1 - \exp(-rR)}{\exp(-rR) - \exp(-r)}\right]. \tag{6.4}$$

For future use, note that the term in brackets is increasing in R.

Substitute the government's budget constraint (6.3) into the individual's one (6.1) to get

$$\int_0^1 c(t)\exp(-rt)dt + \int_R^1 p\exp(-rt)dt = \int_0^R w\exp(-rt)dt.$$

This is the same as (5.1). It is important that this substitution is done only *after* the individual's optimization problem has been solved. This is because the individual does not perceive their own old-age social security as being directly tied to their *own* taxes; that is, the person does not believe that changing the date of retirement, R, will affect the size of the government benefits, s, that they will collect. If this substitution is done before the individual's optimization has been solved, the old first-order condition (5.2) will obtain instead of (6.2). This will be revisited when private savings accounts are discussed. As before, $c(t) = c(0) \exp[(r - \beta)t]$. Plug this into the above equation and integrate to obtain once again

$$\frac{c(0)}{\beta}[1 - \exp(-\beta)] = \frac{w}{r}[1 - \exp(-Rr)] - \frac{p}{r}[\exp(-Rr) - \exp(-r)],$$

which is identical to (5.3). Next, using the fact that $c(0) = 1/\lambda$ in the second first-order condition yields

$$c(0) = \frac{p + (1 - \tau)w - s}{\rho} \exp[-(r - \beta)R].$$

This fact allows the previous equation to be rewritten as

$$\frac{p/w + (1 - \tau) - s/w}{\beta\rho} \exp[-(r - \beta)R][1 - \exp(-\beta)]$$

$$= \frac{1}{r}[1 - \exp(-Rr)] - \frac{p/w}{r}[\exp(-Rr) - \exp(-r)].$$

Last, substitute out for s using (6.4) to obtain

$$\frac{p/w + 1 - \tau\{1 + [1 - \exp(-rR)]/[\exp(-rR) - \exp(-r)]\}}{\beta\rho}$$

$$\times \exp[-(r - \beta)R][1 - \exp(-\beta)]$$

$$= \frac{1}{r}[1 - \exp(-Rr)] - \frac{p/w}{r}[\exp(-Rr) - \exp(-r)],$$

which implicitly provides a solution for the date of retirement, R.

6.2 Analyzing the Retirement Decision, Again

The right-hand side of the above equation is the same as (5.4). The left-hand side is more complicated, but is still decreasing in R—on this, recall that $[1 - \exp(-rR)]/[\exp(-rR) - \exp(-r)]$ is increasing in R. So again, the situation can be portrayed by a diagram similar to figure 5.9. An increase in the tax rate, τ, will reduce the left-hand-side curve, while the right-hand-side one will be unaffected. Thus, the date of retirement will fall.

Proposition 3 (*Tax-Financed Old-Age Social Security*) *An increase in the tax rate for old-age social security, τ, leads to earlier retirement, or a drop in R.*

Intuitively, providing tax-financed old-age social security reduces the cost at the margin of retirement. Since labor income is taxed and a pension is provided in retirement, the gain from working is less. This leads to earlier retirement.

6.3 Private Savings Accounts

Suppose instead that the government mandates that an individual saves the fraction of their income τ, which goes into a private savings account that earns the rate of return r. Now, the individual's date of retirement will affect the size of the benefits that they will collect, because it will have an impact on the balance in their personal savings account. As before, the person's budget constraint is given by (6.1), but now the individual realizes that s is governed by their own actions through (6.3). Substituting (6.3) into (6.1) gives (5.1). This implies that private savings accounts will not distort the person's retirement decision. In fact, in a world where individuals are free to borrow and save (at the same interest rate as the government), private savings accounts will have absolutely no impact on the date of retirement: the government's mandated level of savings can always be undone by borrowing or lending on the private market. (Note that under the current social security system people can borrow against their retirement income. The current social security system distorts the retirement decision, however, as was shown. It taxes work and subsidizes retirement.)

7 Literature Review

The section on better health is inspired by Suen (2006). Braun, Kopecky, and Koreshkova (2017) document how Americans face the risk of ending life being old, sick, and poor. They analyze the impact of social insurance programs designed to mitigate this risk. Health-care spending on humans and pets is discussed by Einav, Finkelstein, and Gupta (2017). The Malawian HIV epidemic is analyzed in Greenwood et al. (2013), which influences section 4. The facts about HIV and the quotation are taken from there. Kremer (1996) was probably the first person to stress the necessity of building behavioral responses into models of AIDS. The material on retirement draws heavily from Kopecky (2011), who develops a model of endogenous retirement. The analysis stresses the importance of falling leisure goods prices and rising wages for explaining the long-run trend toward spending a greater fraction of life retired. In a related vein, recent research by Aguiar et al. (2017) suggests that part of the decline in hours worked by young males between the ages 21 and 30 over the period 2000 to 2015 may be due to innovations in recreational computing and video games. This leisure good is disproportionately favored by young men relative to women and older men. The nondistorting nature of private savings accounts is recognized in Prescott (2004). McGrattan and Prescott (2017) undertake a quantitative analysis of the welfare-improving nature of private savings accounts.

8 Problems

(1) *The Value of Exercise.* Imagine a person lives for two periods. In the first period the individual can exercise, which increases the odds of being fit in the second period. A fit person works in the second period and earns the after-tax wage πw, where π is the individual's productivity level and w is the after-tax wage rate per unit of productivity. An unfit person cannot work, and the government provides such a person with a fixed level of support $c < \pi w$. The individual discounts the future at rate β, where $0 < \beta < 1$. The probability of remaining fit in the second period is p. The person hates exercise, e, and the first-period utility linked with exercising is simply

$-e$.

To attain a fitness probability of p, the person must exercise in the first period in the amount

$e = p^{1+1/\gamma}/(1+1/\gamma)$, with $\gamma > 0$.

The individual's utility function over second-period consumption is

$\ln(c)$,

where c is the person's consumption.

(a) Set up and solve the person's problem for p. (The person does not consume or work in the first period.)

(b) How does exercise, e, vary with the discount factor, β, the wage rate, w, and the level of social support c? What is the economics underlying these results? What is the model's prediction about the level of exercise that a high-wage earner (a high π) will do relative to a low-wage one (a low π)?

(2) *Health Care Spending on Pets.* Imagine the plight of a woman whose dog becomes sick. The vet tells the owner that she can treat the dog at a cost of q. With probability p, the dog will recover and with probability $1-p$, it will die. Without treatment the dog will perish. The woman loves her dog and gets utility from the pet in the amount $\mathfrak{d}$. Should the dog die, either because the treatment fails or because it isn't treated, the owner will suffer grief in the amount $-\mathfrak{g}$. The owner has income in the amount y, which can be used for consumption, c, and to treat the dog. Her utility function for consumption is $\ln(c)$.

(a) Formulate the woman's decision about whether to treat the dog or not.

(b) How is the decision to treat the dog affected by p, $l \equiv \mathfrak{d} + \mathfrak{g}$, and q/y?

(c) Suppose that $l \equiv \mathfrak{d} + \mathfrak{g}$ is distributed across the population in accordance with the cumulative distribution function $H: [0, \infty) \to [0, 1]$. How will an improvement in veterinarian medicine that results in p rising affect the share of aggregate spending on veterinarian care? (Assume that the dog-owning fraction of the population is π and that the probability that a dog gets sick is ϕ. Both of these variables remain constant.) How does this relate to figure 5.4?

(d) Suppose q/y drops instead. Is it possible to say what will happen to the share of aggregate spending on veterinarian care? What is the economics underlying your answer?

6 Conclusion

Relentless tides of technological progress have shaped the household and culture just as waves have shaped the shoreline. This reshaping of the household landscape is reflected in a rise of married female labor-force participation, a decline in fertility, a drop in marriage, changes in culture, and increased longevity and longer retirement.

One of the biggest changes in the twentieth century was the rise in married female labor-force participation. The analysis stressed three factors: labor-saving technological progress in the home, technological advance in the market, and social change that protects women's rights in the workplace. Technological progress in the home was manifested by the introduction of labor-saving household inputs, such as cell phones, dishwashers, dryers, food processors, irons, microwaves, pc's, refrigerators, vacuum cleaners, and washing machines, among other things. The analysis used household production theory to show how the development of such inputs liberated women from the home. Key was the concept of Edgeworth-Pareto substitutability. That is, the advent of such technologies reduced the marginal benefit of labor at home, resulting in a shedding of household labor. This force pushed women into the labor force so to speak.

Technological progress in the market reduced the importance of physical strength relative to mental tasks. This was reflected in a rise of white-collar jobs relative to blue-collar ones. The shift in the nature of jobs from brawn to brain encouraged married women to enter the labor force. As mental skills became more valuable relative to brawn, the wages of jobs at which women have a comparative advantage rose. This worked to reduce the gender wage gap. Last, discrimination against women in the workplace abated, as is discussed below. The workplace then became more enticing for women.

Less gender discrimination eased the friction of moving from working at home to working in the labor market.

Families are much smaller today than in the past. Economics forces can explain the drop in fertility. The inquiry here emphasized the cost of raising children. Children are costly in terms of the parental time involved. With growing real wages, this translates into an ever-increasing cost in terms of foregone consumption, which in turn motivated households to cut back on family size. The rise in the opportunity cost of children was interrupted in the middle of twentieth century. The introduction of labor-saving household technologies reduced the cost of having children. So did progress in obstetric and pediatric medicine, which reduced the illnesses and disabilities associated with bearing children. These forces led to the baby boom. Eventually, the pressure of rising real wages took over again and the decline in fertility resumed.

The time required to bear and raise children influences the type of jobs that women take, in addition to their family size. Taking time off in certain occupations is more costly than in others. The cost of taking time off may be particularly large, in terms of wage penalties, for business and science-related jobs. This may lead to women choosing other more amenable occupations, which are generally lower paying. The gender wage gap reflects these circumstances. As fertility falls, due to ever growing real wages, families will desire fewer children. This makes inflexible high-paying jobs relatively more advantageous than flexible low-paying ones. Thus, women will shift into high-paying occupations. Last, the drop in fertility during World War I France was explained by an increase in the expected cost of children. Households had to worry about whether the man would be maimed or killed in the war, which would lower income and consumption. The potential loss of income and consumption made children more costly.

Technological progress has made it much more affordable for an adult to live alone. Consequently, there has been a decline in the fraction of the population that is married. Marriage offers economies of scale in consumption. The cost of a couple living together is less than the cost of two singles living alone. When deciding whether to live together with someone, a person rationally weighs the utility from single life against the utility from married life. Both economic (money) and noneconomic (love) considerations enter into this comparison. Rising wages and labor-saving technological progress in the home have whittled away the scale economy advantage of married versus single life. As a result, there is less need for people to live together, if they don't really love each other. One might expect that this fading economic motive for marriage has had a bigger impact on people at the lower end of the income distribution as opposed to those at the upper end. Indeed, the decline in marriage and the rise in divorce have been more pronounced for those without a college education than for those with one.

In the U.S. data, there has been an increase in positive assortative mating where people tend to marry or mate with individuals from the same socioeconomic class. As married women migrated from working at home to working in the market, their contribution to household income became much more important. Therefore, the value of a woman's productivity in the market rose relative to the value of her productivity in household production. To the extent that people sort on the basis of economic considerations, this will lead to an increased propensity of college-educated (non-college-educated) men to match with college-educated (non-college-educated) women. Assortative mating will never appear perfect because some characteristics, such as a person's value in the home or the extent of mutual attraction, cannot be directly observed.

Attitudes, culture, laws, and social norms are all functions of technology, too. So, while attitudes, culture, laws, and social norms affect the economy by influencing behavior, the economy affects attitudes, culture, laws, and social norms. Women's rights in the workplace were strengthened in the second half of the twentieth century. This social change in the workplace was illustrated using a median voter model. Households had two considerations when deciding whether or not to vote in favor of women rights. On the one hand, a household values the extra income that a working mother brings home. On the other hand, they may be concerned with a woman's role in the home. In particular, they may have reservations about less time being spent with children, both their own kids and other people's. So, when voting, a household weighs these two considerations against each other. Labor-saving technological progress in the home means that mothers can work without reducing time spent with their children. Additionally, rising wages, in general, and decreases in the gender wage gap, specifically, make it more advantageous for women to work. These trends lead toward a yes vote. At some tipping point, so to speak, the majority of households in society will be in favor of married women working in the workplace.

The sexual revolution is an excellent example of how technology affects culture. Over the course of the twentieth century the failure rate for contraception dropped drastically. This had a profound impact on sexual mores. The fraction of teenage girls with premarital sexual experience rose continuously throughout this period, from almost none at the start of century to nearly three-quarters today. Out-of-wedlock births for teenage girls also rose, although this trend seems to have reversed recently. In the analysis, a teenage girl rationally weighed the costs and benefits of becoming sexually active. Technological progress in contraception dramatically reduced the cost side of the equation, since sex became safer. Whether technological progress in contraception raises or lowers teenage out-of-wedlock births is an elasticity question. In particular, does the decline in the failure rate dominate the rise in the number of sexually active teenagers? When it does, out-of-wedlock births fall, and when it doesn't, they rise.

The social mores and norms inculcated into individuals are also a function of the technological environment. Parents and social institutions mold children's preferences in order to influence their behavior in later life. They do this while recognizing that children, when they grow up, will act in their own best interests; that is, the adult children will maximize their utility subject to the constraints they face. Parental socialization applies to many things, such as not engaging in risky behavior or the value of patience, which involves a sacrifice today for a gain tomorrow. Socializing children is a costly activity, so there are limits on how much is done. In yesteryear out-of-wedlock births were an enormous burden on parents and churches; most families lived close to the subsistence level. Hence, premarital sex was stigmatized so that women would feel a sense of shame if they had an out-of-wedlock birth. As contraception improved, there was less of a need to socialize children about the perils of premarital sex. Society's attitudes toward sex became more liberal.

Life expectancy shot up over the course of the twentieth century. Two factors were stressed here. First is the growth in wages. This increases the value of extending life into the future, since it generates a higher utility value from living and encourages spending on health care. Second is technological advance in medicine. This reduces the cost of extending life and thus promotes health-care spending. Growth in incomes spurs technological progress in medicine. Medical innovation is expensive, and firms need to recover the cost of research and development. Wealthy patients help firms do so. The price of new medical innovations will fall over time, as production becomes more efficient and as competitors enter into the market. As a result, the availability of cutting-edge medical practices will spread to the general population.

In the 1800s the vast majority of people worked all of their lives. Today the vast majority of seniors enjoy a retirement. What brought this about? The analysis here stressed the concept of *leisure goods*. Since leisure goods are time intensive, taking off a block of time to enjoy them, as in retirement, makes sense. Technological progress made leisure goods more affordable due to both rising wages and declining prices. This created an incentive for individuals to take more time off at the end of their lives to enjoy themselves. Social Security was introduced in the United States in 1935. By taxing wages and providing a pension, this reduced the cost of retirement. Therefore, the advent of Social Security could lead to an increase in retirement. It should be noted, however, that the trend in retirement started well before the introduction of U.S. Social Security. Thus, technological progress is probably the primary cause.

It's dangerous to prognosticate about the future. Still, it is safe to say that time spent on housework will decline further. The spread of information technologies such as artificial intelligence into the household and the increased usage of online shopping will reduce the time spent on purchasing household goods and services. Additionally, robotic devices used to clean floors and cut lawns will certainly get better and show price reductions. Ride-sharing services, and eventually driverless cars, will cut the cost of ferrying kids around to activities, lessons, and social events—plus, with GPS tracking parents will know exactly where they are. Additionally, the cost of home delivery will drop with the advent of autonomous vehicles.

As reproductive medicine, such as in vitro fertilization, advances and becomes more accessible, women will have children later in life. By delaying children, women can gain more on-the-job experience in their early working years. Additionally, telecommuting will get better and become more widespread, permitting them to work at home when they have young children. These trends should operate to narrow the gender wage gap.

On marriage and divorce, improvements in online dating services will facilitate better matches between partners. Positive assortative mating might also increase as a result. The implications for marriage and divorce are unclear. On the one hand, better matching ought to reduce divorce.

On the other hand, divorce will be more attractive because it will be easier to find someone else. Social media and the like will allow noncustodial divorced parents to keep in much better contact with their children, so the cost of divorce might not be as great. Additionally, young people may delay marriage because it will be easier to find someone when older.

Last, advances in medicine will certainly continue to prolong life and improve its quality. People will probably spend both more time working and more time in retirement. Retirement will be more enjoyable as the price of leisure goods continues to decline. Future developments in computer gaming, streaming of entertainment and instructional videos, and the like, will cause rises in the value of home leisure. The quality-adjusted prices of such things have dropped rapidly over the past few decades. Additionally, think about the ease of online cost-reducing short-term apartment rentals versus hotel accommodation. This reduces the cost of a vacation, a quintessential leisure good. Diagnostic tools for medicine will enter the home. Along with advances in telecommunication and artificial intelligence, this will result in people transmitting data and interacting with medical service providers in a more effective and efficient manner, providing both better medicine and time savings. Consequently, the elderly will be able to stay at home longer before moving into long-term care. Improvements in video networking will allow the elderly to keep in better touch with family. Maintaining better health will be easier with diet and fitness apps. Thus, technological progress has led to an evolution in the form of households and has made an imprint on everyday life. It will continue to do so in the future, and mostly in ways that are unforeseen.

Mathematical Appendix

Some of the basic mathematics used in the book are reviewed here. This should make the book self-contained for those who are rusty or unfamiliar with the mathematics used. The presentation is cookbook in style and is oriented toward discussing the uses of mathematics in the main text. An excellent gentle and gradual introduction to the mathematics used in economics is contained in Chiang (1984). A good introduction to probability and statistics is DeGroot (1975).

1.A   Maximizing a Function

Maximization is at the heart of economics. Economic actors try to do the best for themselves. So, people maximize their utility, and firms maximize their profits. Mathematically speaking, this corresponds to maximizing a function. Functions are everywhere in economics: they include cost functions, production functions, utility functions to name a few.

Consider the function

$$y = F(x),$$

which maps the real-valued variable x into a real value for the variable y. By definition, a function associates each value of x with a *unique* value for y. Take x to be a nonnegative number, so $x \in \mathcal{R}_+$, and y to be some real number, implying $y \in \mathcal{R}$. Thus, $F : \mathcal{R}_+ \to \mathcal{R}$. Assume that F is continuously twice differentiable. Denote the first and second derivatives of F by

$$F_1(x) \equiv \frac{dF(x)}{dx} \quad \text{and} \quad F_{11}(x) \equiv \frac{d^2 F(x)}{dx^2} = \frac{dF_1(x)}{dx}.$$

The first derivative gives the impact that a small change in x will have on y. The second derivative specifies how the first derivative changes in response to a small shift in x. In other words, it says how the change in y in response to a tiny shift in x, changes with a small movement in x.

Now, consider the unconstrained maximization problem

$$\max_{x} F(x).$$

Here the value of x that maximizes the function $F(x)$ is sought. At a maximum, the following first-order condition must obtain:

$$F_1(x) = 0.$$

This condition is necessary for a local maximum. Suppose to the contrary that at a maximum $F_1(x) > 0$. Then, a small shift up in x would increase $F(x)$, a contradiction. The above first-order condition represents one equation in one unknown, x. The first-order condition specifies a local maximum, instead of a local minimum (or an inflection point), if the second-order condition shown below holds

$$F_{11}(x) < 0.$$

Let x^* denote the value of x that maximizes the the function, $F(x)$. When the second-order condition holds, a small increase in x must cause the function $F(x)$, when evaluated at x^*, to decrease, because $F_1(x)$ becomes negative. Likewise, a small decrease in x induces $F_1(x)$ to become positive, which implies that the reduction in x also results in a decline in $F(x)$. Therefore, x^* must maximize $F(x)$, at least locally.

1.A.1 Strict Concavity (Convexity) and the Second-Order Condition for a Maximum (Minimum)

A strictly concave function has a negative second derivative. That is, if a function is strictly concave, then $F_{11}(x) < 0$ for all x. In this situation, the second-order condition for a maximum will automatically hold; hence, for strictly concave functions the first-order condition is both necessary and sufficient for characterizing a maximum. By contrast, a strictly convex function has a positive second derivative so that $F_{11}(x) > 0$ for all x. In this case, the first-order condition is both necessary and sufficient for a minimum to hold.

This situation is portrayed by figure A.1 for a typical case in economics. At the peak of the function, the slope or the first derivative is zero. Note that the second derivative is negative. That is, the first derivative declines

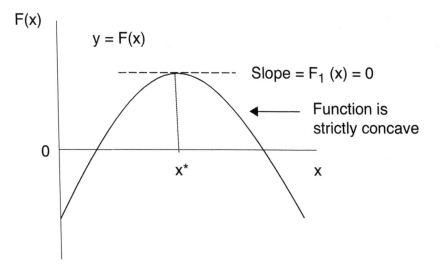

Figure A.1
Finding an unconstrained maximum.

as you move from left to right. This occurs because the objective function is strictly concave in x.

1.A.2   Corner Solutions

In economics corner solutions to maximization problems often occur. For example, perhaps a person wants to set their hours worked in the labor force to be zero, or likewise, they do not want to acquire any skill by attaining a post-secondary education. Now, suppose that there is a lower bound on x, denoted by x_l, so that the constraint $x \geq x_l$ must hold. The maximization problem above now appears as

$$\max_{x \geq x_l} F(x).$$

One of two solutions may obtain to this constrained maximization problem: an interior solution or a corner solution. The interior solution is described as before by the first-order condition

$$F_1(x) = 0.$$

The corner solution occurs when

$$F_1(x_l) < 0.$$

This is shown by the bottom panel of figure A.2. Here, the peak of the function cannot be attained because the lower bound on x has been hit. The slope is negative at $x = x_l$. Because $F_1(x) < 0$, at $x = x_l$, a small reduction in x would increase the value of the objective function, $F(x)$. This cannot be done due to the presence of the lower bound, x_l. Alternatively, x could be constrained by an upper bound, x_u, which requires $x \leq x_u$. Now the corner solution happens when

$$F_1(x_u) > 0.$$

A small increase in x from x_u would raise the value of the objective function, but this isn't feasible, because the upper bound, x_u, has been hit. The top panel of figure A.2 illustrates this situation.

1.A.3 Constrained Maximization

Optimization in economics often involves maximization subject to constraints. For example, a consumer maximizes utility subject to a budget constraint, or a firm minimizes costs subject to a production function. Consider maximizing the function $F(x, y)$, with respect to the decision variables x and y, subject to a constraint given by the function $y = G(x)$. There are two ways to proceed here. First, one could just replace y in the objective function with the function $G(x)$ and then maximize with respect to the single variable, x. That is, one could solve the problem

$$\max_x F(x, G(x)).$$

By defining the new function $\tilde{F}(x) = F(x, G(x))$, this reduces to the form of the maximization problem discussed above. All but one of the constrained maximization problems in this book can be handled in this elementary way.

Second, the method of Lagrange multipliers can be used. To do this, set up the Lagrangian function

$$L = F(x, y) + \lambda [G(x) - y].$$

The variable λ is called the *Lagrange multiplier*. The term in brackets is just the constraint. Now, take the derivatives with respect to x, y, and λ, and set them to 0, to get

$$\frac{dL}{dx} = F_x(x, y) + \lambda G_x(x) = 0, \tag{1.A.1}$$

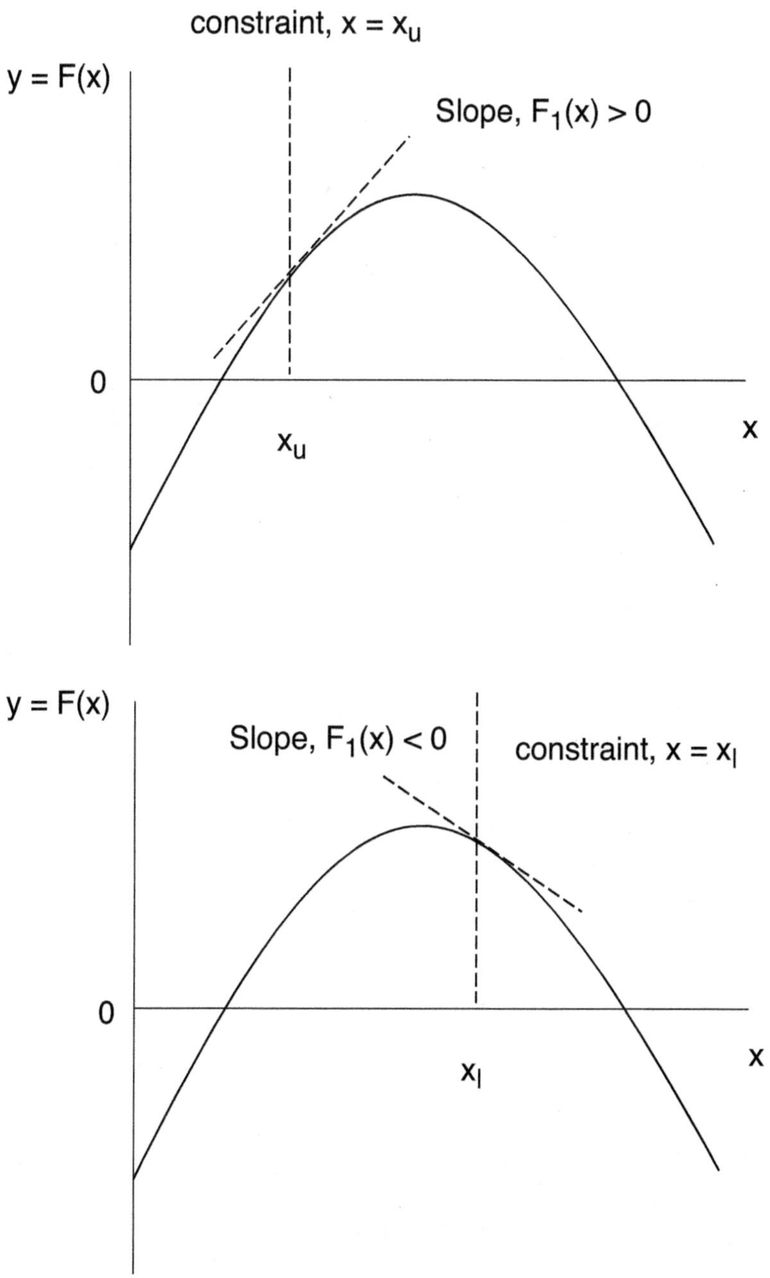

Figure A.2
Constrained maximization. The bottom panel shows the situation when a corner solution is hit at a lower bound, while the top one illustrates things for an upper bound.

$$\frac{dL}{dy} = F_y(x, y) - \lambda = 0, \tag{1.A.2}$$

$$\frac{dL}{d\lambda} = G(x) - y = 0,$$

where $F_x(x, y) \equiv dF(x, y)/dx$, $F_y(x, y) \equiv dF(x, y)/dy$, and $G_x(x) \equiv dG(x)/dx$. The last equation just gives back the constraint. The above implicitly represents a system of three nonlinear equations in the three unknowns, x, y, and λ. It defines a solution to the constrained maximization problem.

1.A.3.1 The Lagrange multiplier as a shadow value

The Lagrange multiplier, λ, has a nice economic interpretation. It gives the shadow value of relaxing the constraint slightly. To see this, rewrite the constraint as $y = G(x) + a$, where a is some exogenous constant. (By setting $a = 0$ the original constraint obtains.) Let x^*, y^*, and λ^* be the solution to the above system of equations, where the last equation has been replaced by

$$\frac{dL}{d\lambda} = G(x) + a - y = 0. \tag{1.A.3}$$

Next, consider the optimized Lagrangian

$$L = F(x^*, y^*) + \lambda^*[G(x^*) + a - y^*].$$

Finally, differentiate L with respect to a to obtain

$$\frac{dL}{da} = \underbrace{[F_x(x^*, y^*) + \lambda^* G_x(x^*)]}_{=0} \frac{dx^*}{da}$$

$$+ \underbrace{[F_y(x^*, y^*) - \lambda^*]}_{=0} \frac{dy^*}{da}$$

$$+ \underbrace{[G(x^*) + a - y^*]}_{=0} \frac{d\lambda^*}{da}$$

$$+ \lambda^*.$$

Observe that the impact of any induced changes in x^*, y^*, and λ^* cancels out due to equations (1.A.1), (1.A.2), and (1.A.3). This is an application of the envelope theorem discussed in Chapter 1. Therefore,

$$\frac{dL}{da} = \lambda^*.$$

This implies that relaxing the constraint by slightly increasing a moves up the Lagrangian by the shadow value of the constraint or λ^*. Thus, λ^* gives the worth of a marginal shift up in a.

2.A   Total Differentials

Consider the function

$$z = F(x, y).$$

What would happen to z if both x and y are changed by some arbitrary small amounts? Denote the small changes in x and y by dx and dy, respectively. These are called *differentials*. Likewise, the induced total change in z is represented by dz. The total change in z is given by

$$dz = F_x(x, y)dx + F_y(x, y)dy, \qquad (2.A.1)$$

where $F_x(x, y) \equiv dF(x, y)/dx$ and $F_y(x, y) \equiv dF(x, y)/dy$. The above expression decomposes the change in z into two factors. The first term on the right-hand side is the change in z that results from the shift in x. The shift in x is represented by dx. To get the induced shift in z, dx is multiplied by the (partial) derivative $F_x(x, y)$, which translates a shift in x into a shift in z. The second term does the same thing for y.

2.A.1   The Total Derivative

The differentials dz, dx, and dy can be manipulated to obtain derivatives. For example, one could divide the above equation through by dx to obtain

$$\frac{dz}{dx} = F_x(x, y) + F_y(x, y)\frac{dy}{dx}.$$

The term dz/dx is the total derivative of z with respect to x. The change in x has both a direct and indirect effect on z, as shown by the first and second terms on the right-hand side. The indirect effect occurs because the change in x may induce a change in y, as given by dy/dx, which in turn will affect z via $F_y(x, y)$. To compute the indirect effect, more information is needed. To illustrate, perhaps y is given by the function $y = G(x)$. Then, $dy/dx = G_x(x)$, implying $dz/dx = F_x(x, y) + F_y(x, y)G_x(x)$. Alternatively, perhaps y is not a function of x. Then, $dy/dx = F_x(x, y)$. To take another example, perhaps x and y are functions of another variable t, which changed by the small amount, dt. Then, dividing both sides of (2.A.1) by dt gives

$$\frac{dz}{dt} = F_x(x, y)\frac{dx}{dt} + F_y(x, y)\frac{dy}{dt}.$$

Here, the derivatives dx/dt and dy/dt would depend on the specified functional dependencies of x and y on t.

3.A Leibniz's Rule

Consider an integral of the form

$$I(a) = \int_{L(a)}^{U(a)} F(a, x)dx,$$

where $F(a, x)$, $L(a)$, and $U(a)$ are all functions. *Leibniz's rule* specifies how this integral can be differentiated with respect to the parameter a. It states that

$$\frac{dI(a)}{da} = F(a, U(a))U_a(a) - F(a, L(a))L_a(a) + \int_{L(a)}^{U(a)} F_a(a, x)dx,$$

where

$$\frac{dF(a, x)}{da} \equiv F_a(a, x), \quad \frac{dL(a)}{da} \equiv L_a(a), \text{ and } \frac{dU(a)}{da} \equiv U_a(a).$$

4.A Distribution Functions and Correlation Coefficients

Cumulative distribution functions (cdf's) play an important role in the book. They are used to characterize how some trait is dispersed across the population at large. Let $\tilde{x}$ be a real-valued random variable. The cumulative distribution function, $G(x)$, gives the probability that $\tilde{x}$ is no greater than x; that is,

$$\Pr[\tilde{x} \le x] = G(x).$$

Distribution functions are always increasing, because it must be the case that $\Pr[\tilde{x} \le x] \le \Pr[\tilde{x} \le x']$ for $x < x'$. This is intuitive in that the statement that $\tilde{x} \le x'$ includes the possibility that $\tilde{x} \le x$, since $x < x'$. Hence, the likelihood that $\tilde{x} \le x$ must be (weakly) smaller than the one that $\tilde{x} \le x'$. It is easy to see that

$$\Pr[\tilde{x} > x] = 1 - G(x).$$

The derivative of the distribution function is the probability density function, $g(x)$, where by definition $g(x) = G_1(x)$. Thus,

$$G(x) = \int^x g(\tilde{x})d\tilde{x}.$$

When looking at some large population, the distribution function gives the fraction of the population that has a value of $\tilde{x}$ below x. For example the fraction of U.S. males, aged 20–29, that is less than or equal to six feet tall is approximately 80 percent, which given the large size of the population is pretty much known with certainty. The probability density function describes how the fraction of the population with an $\tilde{x} \leq x$ changes with x.

4.A.1   Three Distribution Functions

Three types of distribution functions are used in this book: the *Pareto distribution*, the *exponential distribution*, and the *uniform distribution*. Their definitions are provided now.

Definition 1 (*Pareto Distribution*)   *A random variable $\tilde{x}$ is distributed according to a Pareto distribution $P : [\psi, \infty) \to [0, 1]$ if*

$$\Pr[\tilde{x} \leq x] = P(x) = 1 - \left(\frac{\psi}{x}\right)^\gamma, \quad \text{where } \psi > 0 \quad \text{and} \quad \gamma > 1,$$

for $x \geq \psi$. The constant ψ is called the scale parameter *and represents the lower bound for the random variable $\tilde{x}$. The constant γ is dubbed the* shape parameter. *The Pareto distribution is strictly increasing and strictly concave in x. The higher γ is, the larger $P(x)$ is for a given x. So, the Pareto distribution shifts up (at every point but $x = \psi$) with higher values for γ. Strictly speaking, the scale parameter just needs to be positive ($\gamma > 0$), but when $0 < \gamma \leq 1$, the mean of Pareto distribution is infinite. By differentiating $P(x)$ with respect to x, is easy to calculate that the associated probability density function is*

$$P_1(x) = \gamma \psi^\gamma \left(\frac{1}{x}\right)^{\gamma+1}.$$

By examining the ratio of $P_1(x)/[1 - P(x)] = \gamma/x$, one can deduce that the proportional rate of change in the fraction of population with an $\tilde{x} > x$, with respect to *x, increases with γ. The mean of x is given by*

$$\int_\psi^\infty x\gamma\psi^\gamma \left(\frac{1}{x}\right)^{\gamma+1} dx = \gamma\psi^\gamma \int_\psi^\infty \left(\frac{1}{x}\right)^\gamma dx$$

$$= \frac{\gamma}{1-\gamma}\psi^\gamma x^{1-\gamma} |_\psi^\infty = \frac{\gamma}{\gamma-1}\psi.$$

Vilfredo Pareto (1897) used the function P to estimate the distribution of income. Suppose that x represents a certain income level. Let n represent the fraction of people with an income level greater than x. Thus, $n = (\psi/x)^\gamma$. This implies $\ln n = \gamma \ln \psi - \gamma \ln x$. This relationship can be estimated using a simple linear regression of the form $\ln n = c + b \ln x$, where $c = \gamma \ln \psi$ and $b = -\gamma$. Pareto examined this relationship over 100 years ago.

Definition 2 (*Exponential distribution*) *A random variable $\tilde{x}$ is distributed according to an exponential distribution $E : [0, \infty) \to [0, 1]$ if*

$$\Pr[\tilde{x} \leq x] = E(x) = 1 - \exp(-x/\beta), \quad \text{where } \beta > 0.$$

The exponential distribution is both strictly increasing and strictly concave in x. The constant β is called the rate parameter *and the bigger it is the larger $E(x)$ is for any given x (except for $x = 0$). It is straightforward to verify that the probability density function is*

$$E_1(x) = \frac{\exp(-x/\beta)}{\beta}.$$

It is easy to see that $E_1(x)/[1 - E(x)] = 1/\beta$, so that β governs the proportional rate of change in the fraction of population with an $\tilde{x} > x$, with respect to x. The mean of x is given by

$$\int_0^\infty \frac{x\exp(-x/\beta)}{\beta}dx = \frac{1}{\beta}\int_0^\infty x\exp(-x/\beta)dx$$

$$= -e^{-\frac{x}{\beta}}(x+\beta) |_0^\infty = \beta.$$

Definition 3 (*Uniform distribution*) *A random variable $\tilde{x}$ is distributed according to a uniform distribution $U : [\underline{x}, \overline{x}] \to [0, 1]$ if*

$$\Pr[\tilde{x} \leq x] = U(x) = \frac{x - \underline{x}}{\overline{x} - \underline{x}},$$

for $\underline{x} \leq x \leq \overline{x}$. The uniform distribution $U(x)$ is just a straight line that starts at $U = 0$ when $x = \underline{x}$ and ends at $U = 1$ when $x = \overline{x}$. The probability density function connected with the uniform distribution is

$$U_1(x) = \frac{1}{\overline{x} - \underline{x}}.$$

The mean of x is given by

$$\int_{\underline{x}}^{\overline{x}} \frac{x}{\overline{x} - \underline{x}} dx = \frac{1}{\overline{x} - \underline{x}} \int_{\underline{x}}^{\overline{x}} x dx$$

$$= \frac{1}{\overline{x} - \underline{x}} \frac{1}{2} x^2 \Big|_{\underline{x}}^{\overline{x}} = \frac{1}{2} \frac{\overline{x}^2 - \underline{x}^2}{\overline{x} - \underline{x}} = \frac{1}{2} \frac{(\overline{x} + \underline{x})(\overline{x} - \underline{x})}{\overline{x} - \underline{x}}$$

$$= \frac{\overline{x} + \underline{x}}{2}.$$

It is easy to deduce that the median is also given by $(\overline{x} + \underline{x})/2$. The variance of x is $(\overline{x} - \underline{x})^2/12$.

4.A.2   Correlation Coefficients

Correlation coefficients measure the association between two data series, say $\{x_i\}_{i=1}^n$ and $\{y_i\}_{i=1}^n$. A correlation coefficient takes a value between -1 and 1, where a positive value indicates that two series tend to move together while a negative one shows that they have a proclivity to move opposite to one another. The higher the correlation coefficient is in absolute value, the stronger is the association. A value of 0 shows no association. Two measures of correlation are used in the book: the *Pearson correlation coefficient* and the *Kendall rank correlation coefficient*. The former measures the degree of linear association between two series, while the latter measures the strength of association in ordinal terms.

Definition 4 (*Pearson correlation coefficient*, ρ)   *The Pearson correlation coefficient is a measure of the linear dependence (or correlation) between two data series, $\{x_i\}_{i=1}^n$ and $\{y_i\}_{i=1}^n$. It is defined by the formula*

$$\rho = \frac{\sum_i^n (x_i - \overline{x})(y_i - \overline{y})}{\sqrt{\sum_i^n (x_i - \overline{x})^2} \sqrt{\sum_i^n (y_i - \overline{y})^2}},$$

where the sample means, $\overline{x}$ and $\overline{y}$, are given by $\overline{x} = (\sum_i^n x_i)/n$ and $\overline{y} = (\sum_i^n y_i)/n$. If there is a tendency when x rises above its mean for y to do so as well, then the numerator will likely be positive and hence so will be ρ. The opposite will be *true, if there is a penchant for y to fall below its mean when x rises above its one. If y and x have a strictly positive (negative) linear relationship then ρ will be 1 (−1).*

Figure A.3 illustrates the Pearson correlation coefficient between x and y for some randomly generated series. The two series for x and y are always positively associated. As can be seen, as ρ increases so does the strength of the positive association.

Definition 5 (Kendall rank correlation coefficient, τ) *Kendall's τ is a measure of the degree of order between two data series, $\{x_i\}_{i=1}^n$ and $\{y_i\}_{i=1}^n$. Consider the case where the two variables are positively associated with each other. Count the number of times when $x_j > x_i$ and $y_j > y_i$. Call the number of times the series are concordant, C. There are a total of $n(n-1)$ comparisons to do. Kendall's τ is given by*

$$\tau = \frac{C - [n(n-1) - C]}{n(n-1)}.$$

The term $n(n-1) - C$ measures the number of times the series are discordant. Essentially, Kendall's τ is the probability of concordance less the probability of discordance. It is easy to see that $-1 \leq \tau \leq 1$. The sign and strength of the relationship are shown by τ. If τ is positive, then the series tend to move together, while if τ is negative, they are inclined to move in opposite fashion. A high absolute value shows that the series are concordant, a low value that they are discordant, and a zero value that they are independent.

5.A A Linear First-Order Difference Equation

Difference equations are common in economics, especially macroeconomics. They characterize the evolution of variables such as GDP over time. A typical linear first-order difference equation in macroeconomics is the following:

$y' = a + by$, with $0 < b < 1$,

where y denotes the current value of the endogenous variable, y' represents its value next period, y'' its value two periods ahead in time, and so on. The restriction that $0 < b < 1$ implies that

(1) The difference equation is stable or that it will converge to a long-run value for y, denoted by y^*;

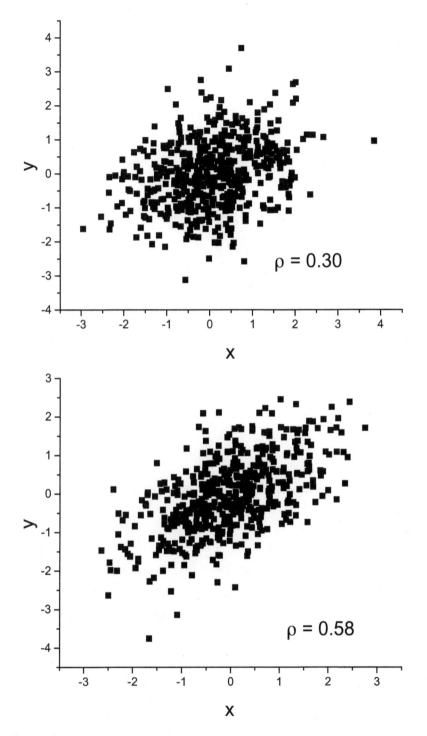

Figure A.3
Pearson correlation coefficient. As one moves from the first to the third diagram, the degree of positive association between the series for *x* and *y* increases. This is reflected in higher values for the Pearson correlation coefficient, ρ.

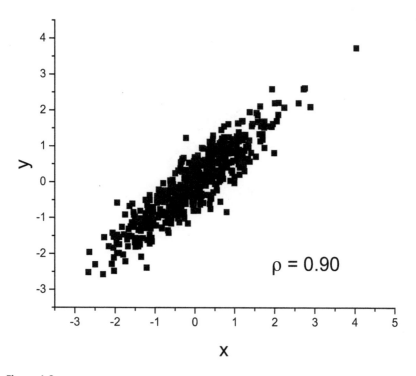

Figure A.3
Continued

(2) Convergence to y^* will be monotone so that if the equation starts off from a value for $y < y^*$, it will rise over time in a strictly increasing fashion to y^*; i.e., if $y < y^*$, then $y < y' < y'' < \cdots \leq y^*$. If $y > y^*$, then $y > y' > y'' > \cdots \geq y^*$.

Now, consider the nonlinear first-order difference equation

$\widetilde{y}' = A\widetilde{y}^{b}$, with $A > 0$ and $0 < b < 1$.

This can be converted to a linear first-order difference equation by taking logarithms of both sides to get

$\ln \widetilde{y}' = \ln A + b \ln \widetilde{y}$.

This can be rewritten as

$y' = a + by$,

by making the following definitions: $y' \equiv \ln \widetilde{y}'$, $y \equiv \ln \widetilde{y}$, and $a \equiv \ln A$.

5.A.1 The Steady-State Solution

In a steady state the variable y does not change over time. Hence, $y = y' = y'' = \cdots = y^*$. To compute the steady-state value for y, just set $y = y' = y^*$ and then solve for y^*. One obtains

$$y^* = \frac{a}{1-b}.$$

5.A.2 Monotone Convergence to the Steady State

Suppose that $y < y^*$. This is true if and only if $y' > y$. To see this, observe that $y' - y = a - (1-b)y$. Therefore, $y' - y > 0$, if and only if $a - (1-b)y > 0$. But, $a - (1-b)y > 0$ in turn, if and only if $y < a/(1-b) = y^*$. Likewise, $y > y^*$, if and only if $y' < y$. Hence,

$$y' \gtrless y, \text{ if and only if } y \lessgtr y^*.$$

Next, assume $y < y^*$. This transpires if and only if $y' < y^*$. This is easy to demonstrate since $y' = a + by < a + by^*$, if and only if $y < y^*$. Now, $a + by^* = y^*$, because y^* is the steady-state solution for y. Therefore, $y' = a + by < y^*$, if and only if $y < y^*$. Similarly, $y > y^*$, if and only if $y' > y^*$. Therefore,

$$y' \lessgtr y^*, \text{ if and only if } y \lessgtr y^*.$$

The above two conditions imply that if $y < y^*$, then y will rise in a monotonic fashion to y^*, and if $y > y^*$, then y will fall in a monotonic fashion to y^*.

References

Abma, Joyce C., Gladys M. Martinez, William D. Mosher, and Brittany S. Dawson. 2004. "Teenagers in the United States: Sexual Activity, Contraceptive Use, and Childbearing, 2002." Vital and Health Statistics, Series 23, no. 24. Hyattsville, MD: National Center for Health Statistics.

Aguiar, Mark, Mark Bils, Kerwin Kofi Charles, and Erik Hurst. 2017. "Leisure Luxuries and the Labor Supply of Young Men." Working Paper No. 23552. Cambridge, MA: National Bureau of Economic Research.

Aguiar, Mark, and Erik Hurst. 2007. "Measuring Trends in Leisure: The Allocation of Time over Five Decades." *Quarterly Journal of Economics* 122 (3): 969–1006.

Aiyagari, S. Rao, Jeremy Greenwood, and Nezih Guner. 2000. "On the State of the Union." *Journal of Political Economy* 108 (2): 213–244.

Albanesi, Stefania, and Claudia Olivetti. 2014. "Maternal Health and the Baby Boom." *Quantitative Economics* 5 (2): 225–269.

Albanesi, Stefania, and Claudia Olivetti. 2016. "Gender Roles and Medical Progress." *Journal of Political Economy* 124 (3): 650–695.

Becker, Gary S. 1965. "A Theory of the Allocation of Time." *Economic Journal* 75 (299): 493–527.

Becker, Gary S. 1971. *The Economics of Discrimination*, 2nd ed. Chicago: University of Chicago Press.

Becker, Gary S. 1973. "A Theory of Marriage: Part I." *Journal of Political Economy* 81 (4): 813–846.

Becker, Gary S., and Robert J. Barro. 1988. "A Reformulation of the Economic Theory of Fertility." *Quarterly Journal of Economics* 103 (1): 1–25.

Becker, Gary S., and Casey B. Mulligan. 1997. "The Endogenous Determination of Time Preference." *Quarterly Journal of Economics* 113 (3): 729–758.

Benhabib, Jess, Richard Rogerson, and Randall Wright. 1991. "Homework in Macroeconomics: Household Production and Aggregate Fluctuations." *Journal of Political Economy* 99 (6): 1166–1187.

Bennett, Judith M. 2003. "Writing Fornication: Medieval Leyrwite and its Historians." *Transactions of the Royal Historical Society* Sixth Series 13: 131–162.

Bethencourt, Carlos, and Jose-Victor Rios-Rull. 2009. "On the Living Arrangements of Elderly Widows." *International Economic Review* 50 (3): 773–801.

Bick, Alexander, Nicola Fuchs-Schundeln, and David Lagakos. 2018. "How Do Hours Worked Vary with Income? Cross-Country Evidence and Implications." *American Economic Review* 108 (1): 170–199.

Bisin, Alberto, and Thierry Verdier. 2011. "The Economics of Cultural Transmission and Socialization." In *Handbook of Social Economics*, vol. 1, edited by Jess Benhabib, Alberto Bisin, and Matthew O. Jackson. The Netherlands: North-Holland, 339–416.

Black, Dan A., Seth G. Sanders, and Lowell J. Taylor. 2007. "The Economics of Lesbian and Gay Families." *Journal of Economic Perspectives* 21 (2): 53–70.

Blau, Francine D., Marianne A. Ferber, and Anne E. Winkler. 2014. *The Economics of Women, Men, and Work*, 7th ed. Boston: Pearson.

Blau, Francine D., and Lawrence M. Kahn. 2008. "Women's Work and Wages." In *The New Palgrave: A Dictionary of Economics*, 2nd ed., edited by Steven N. Durlauf and Lawrence E. Blume. London: Palgrave Macmillan, online.

Blau, Francine D., and Lawrence M. Kahn. 2017. "The Gender Wage Gap: Extent, Trends, and Explanations." *Journal of Economic Literature* 55 (3): 789–865.

Boldrin, Michele, Mariacristina de Nardi, and Larry E. Jones. 2015. "Fertility and Social Security." *Journal of Demographic Economics* 81 (3): 261–299.

Braun, R. Anton, Karen A. Kopecky, and Tatyana Koreshkova. 2017. "Old, Sick, Alone, and Poor: A Welfare Analysis of Old-Age Social Insurance Programmes." *Review of Economic Studies* 84 (2): 580–612.

Bridgman, Benjamin, Georg Duernecker, and Berthold Herrendorf. 2017. "Structural Transformation, Marketization, and Household Production around the World." Unpublished paper, Bureau of Economic Analysis, U.S. Department of Commerce.

Bronson, Mary Ann. 2015. "Degrees Are Forever: Marriage, Educational Investment, and Lifecycle Labor Decisions of Men and Women." Unpublished paper, Department of Economics, Georgetown University.

Browning, Martin, Pierre-Andre Chiappori, and Yoram Weiss. 2014. *Economics of the Family*. New York: Cambridge University Press.

Cain, Glenn G. 1966. *Married Women in the Labor Force: An Economic Analysis.* Chicago: University of Chicago Press.

Caplow, Theodore, Louis Hicks, and Ben J. Wattenberg. 2001. *The First Measured Century: An Illustrated Guide to Trends in America, 1900–2000.* Washington, DC: AEI Press.

Caucutt, Elizabeth, Nezih Guner, and John Knowles. 2002. "Why Do Women Wait? Matching, Wage Inequality, and the Incentives for Fertility Delay." *Review of Economic Dynamics* 5 (4): 815–855.

Caucutt, Elizabeth, Nezih Guner, and Christopher Rauh. 2016. "Is Marriage for White People? Incarceration and the Racial Marriage Divide." Unpublished paper, Department of Economics, Western University.

Cavalcanti, Tiago V. de V., Georgi Kocharkov, and Cezar Santos. 2017. "Family Planning and Development: Aggregate Effects of Contraceptive Use." REAP Working Paper 105, Rede de Economia Aplicada, Brazil.

Cavalcanti, Tiago V. de V., and Jose Tavares. 2008. "Assessing the 'Engines of Liberation': Home Appliances and Female Labor Force Participation." *Review of Economics and Statistics* 90 (1): 81–88.

Cavalcanti, Tiago V. de V., and Jose Tavares. 2011. "Women Prefer Larger Governments: Growth, Structural Transformation and Government Size." *Economic Inquiry* 49 (1): 155–171.

Chiang, Alpha C. 1984. *Fundamental Methods of Mathematical Economics*, 3rd ed. New York: McGraw-Hill.

Chiappori, Pierre-Andre, Murat F. Iyigun, and Yoram Weiss. 2009. "Investment in Schooling and the Marriage Market." *American Economic Review* 99 (5): 1689–1713.

Chiappori, Pierre-Andre, Sonia Oreffice, and Climent Quintana-Domeque. 2012. "Fatter Attraction: Anthropometric and Socioeconomic Matching on the Marriage Market." *Journal of Political Economy* 120 (4): 659–695.

Clark, Gregory. 2010. "The Macroeconomic Aggregates for England, 1209–2008." In *Research in Economic History*, vol. 27, edited by Alexander J. Field. Bingley, UK: Emerald Group Publishing, 51–140.

Coen-Pirani, Daniele, Alexis Leon, and Steven Lugauer. 2010. "The Effect of Household Appliances on Female Labor Force Participation: Evidence from Micro Data." *Labour Economics* 17 (3): 503–513.

Cortes, Guido Matias, Nir Jaimovich, and Henry E. Siu. 2017. "The 'End of Men' and Rise of Women in the High-Skilled Labor Market." Unpublished paper, Vancouver School of Economics, University of British Columbia.

DeGroot, Morris H. 1975. *Probability and Statistics*. Reading, MA: Addison-Wesley Publishing.

Dinkelman, Taryn. 2011. "The Effects of Rural Electrification on Employment: New Evidence from South Africa." *American Economic Review* 101 (7): 3078–3108.

Doepke, Matthias. 2004. "Accounting for Fertility Decline during the Transition to Growth." *Journal of Economic Growth* 9 (3): 347–383.

Doepke, Matthias. 2005. "Child Mortality and Fertility Decline: Does the Barro-Becker Model Fit the Facts?" *Journal of Population Economics* 18 (2): 337–366.

Doepke, Matthias, and Michele Tertilt. 2009. "Women's Liberation: What's in It for Men?" *Quarterly Journal of Economics* 124 (4): 1541–1591.

Doepke, Matthias, and Fabrizio Zilibotti. 2008. "Occupational Choice and the Spirit of Capitalism." *Quarterly Journal of Economics* 123 (2): 747–793.

Easterlin, Richard A. 1987. "Easterlin Hypothesis." In *The New Palgrave: A Dictionary of Economics*, vol. 2, edited by John Eatwell, Murray Milgate, and Peter Newman. London: Macmillan, 1–4.

Eckstein, Zvi, Pedro Mira, and Kenneth I. Wolpin. 1999. "A Quantitative Analysis of Swedish Fertility Dynamics: 1751–1990." *Review of Economic Dynamics* 2 (1): 137–165.

Einav, Liran, Amy Finkelstein, and Atul Gupta. 2017. "Is American Pet Health Care (also) Uniquely Inefficient?" *American Economic Review* 107 (5): 491–495.

Ermisch, John F. 2003. *An Economic Analysis of the Family*. Princeton, NJ: Princeton University Press.

Fernandez, Raquel. 2011. "Does Culture Matter?" In *Handbook of Social Economics*, vol. 1, edited by Jess Benhabib, Alberto Bisin, and Matthew O. Jackson. The Netherlands: North-Holland, 482–510.

Fernandez, Raquel, Alessandra Fogli, and Claudia Olivetti. 2004. "Mothers and Sons: Preference Formation and Female Labor Force Dynamics." *Quarterly Journal of Economics* 119 (4): 1249–1299.

Fernandez, Raquel, and Joyce Cheng Wong. 2017. "Free to Leave? A Welfare Analysis of Divorce Regimes." *American Economic Journal: Macroeconomics* 9 (3): 72–115.

Fernandez-Villaverde, Jesus, Jeremy Greenwood, and Nezih Guner. 2014. "From Shame to Game in One Hundred Years: An Economic Model of the Rise in Premarital Sex and Its De-Stigmatization." *Journal of the European Economic Association* 12 (1): 25–61.

Frederick, Christine. 1912. "The New Housekeeping: How It Helps the Woman Who Does Her Own Work." *Ladies Home Journal* 13 (September): 13 and 70–71.

Fuchs, Rachel G. 1992. *Poor and Pregnant in Paris: Strategies for Survival in the Nineteenth Century*. New Brunswick, NJ: Rutgers University Press.

Gale, David, and Lloyd S. Shapley. 1962. "College Admissions and the Stability of Marriage." *American Mathematical Monthly* 69 (1): 9–14.

Galor, Oded, and David N. Weil. 1996. "The Gender Gap, Fertility, and Growth." *American Economic Review* 86 (3): 374–387.

Galor, Oded, and David N. Weil. 2000. "Population, Technology, and Growth: From Malthusian Stagnation to the Demographic Transition and Beyond." *American Economic Review* 90 (4): 806–828.

Gayle, George-Levi, Limor Golan, and Mehmet A. Soytas. 2015. "What Accounts for the Racial Gap in Time Allocation and Intergenerational Transmission of Human Capital?" Unpublished paper, Department of Economics, Washington University.

Gemici, Ahu, and Steven Laufer. 2010. "Marriage and Cohabitation." Unpublished paper, Department of Economics, Royal Holloway, University of London.

Gershoni, Naomi, and Corinne Low, 2017. "The Impact of Extended Reproductive Time Horizons: Evidence from Israel's Expansion of Access to IVF." Unpublished paper, The Wharton School, University of Pennsylvania.

Gershuny, Jonathan, and Teresa Attracta Harms. 2016. "Housework Now Takes Much Less Time: 85 Years of US Rural Women's Time Use." *Social Forces* 95 (2): 503–524.

Giedion, Siegfried. 1948. *Mechanization Takes Command: A Contribution to Anonymous History*. New York: Oxford University Press.

Godbeer, Richard. 2002. *Sexual Revolution in Early America*. Baltimore: Johns Hopkins University Press.

Goldin, Claudia. 1990. *Understanding the Gender Gap: An Economic History of American Women*. New York: Oxford University Press.

Greenwood, Jeremy, and Nezih Guner. 2009. "Marriage and Divorce since World War II: Analyzing the Role of Technological Progress on the Formation of Households." *NBER Macroeconomics Annual 2008* 23: 231–276.

Greenwood, Jeremy, and Nezih Guner. 2010. "Social Change: The Sexual Revolution." *International Economic Review* 51 (4): 893–923.

Greenwood, Jeremy, Nezih Guner, Georgi Kocharkov, and Cezar Santos. 2016. "Technology and the Changing Family: A Unified Model of Marriage, Divorce, Educational Attainment, and Married Female Labor-Force Participation." *American Economic Journal: Macroeconomics* 8 (1): 1–41.

Greenwood, J., N. Guner, and G. Vandenbroucke. 2017. "Family Economics Writ Large." *Journal of Economic Literature* 55 (4): 1346–1434.

Greenwood, Jeremy, and Zvi Hercowitz. 1991. "The Allocation of Capital and Time over the Business Cycle." *Journal of Political Economy* 99 (6): 1188–1214.

Greenwood, Jeremy, Philipp Kircher, Cezar Santos, and Michele Tertilt. 2013. "An Equilibrium Model of the African HIV/AIDS Epidemic." Unpublished paper, Department of Economics, University of Mannheim.

Greenwood, Jeremy, and Ananth Seshadri. 2002. "The U.S. Demographic Transition." *American Economic Review* 92 (2): 153–159.

Greenwood, Jeremy, Ananth Seshadri, and Guillaume Vandenbroucke. 2005. "The Baby Boom and Baby Bust." *American Economic Review* 95 (1): 183–207.

Greenwood, Jeremy, Ananth Seshadri, and Mehmet Yorukoglu. 2005. "Engines of Liberation." *Review of Economic Studies* 72 (1): 109–133.

Greenwood, Jeremy, and Gokce Uysal. 2005. "New Goods and the Transition to a New Economy." *Journal Economic Growth* 10 (2): 99–134.

Gronau, Reuben. 1986. "Home Production—A Survey." In *Handbook of Labor Economics*, vol. 1, edited by Orley C. Ashenfelter and Richard Layard. Amsterdam: Elsevier, 274–304.

Hansen, Gary D., and Edward C. Prescott. 2002. "Malthus to Solow." *American Economic Review* 92 (4): 1205–1217.

Hayden, Mary. 1942–1943. "Charity Children in Eighteenth-Century Dublin." *Dublin Historical Record* 5: 92–107.

Hazan, Moshe, David Weiss, and Hosny Zoabi. 2017. "Women's Liberation as a Financial Innovation." *Journal of Finance*, forthcoming.

Heisig, Jan Paul. 2001. "Who Does More Housework: Rich or Poor? A Comparison of 33 Countries." *American Sociological Review* 76 (1): 74–99.

Heston, Alan, and Robert Summers. 1991. "The Penn World Table (Mark 5): An Expanded Set of International Comparisons, 1950–1980." *Quarterly Journal of Economics* 106 (21): 327–368.

Himes, Norman E. 1963. *Medical History of Contraception*. New York: Gamut Press.

Hyslop, Dean R. 2001. "Rising U.S. Earnings Inequality and Family Labor Supply: The Covariance Structure of Intrafamily Earnings." *American Economic Review* 91 (4): 755–777.

Jones, Larry E., Rodolfo E. Manuelli, and Ellen R. McGrattan. 2015. "Why Are Married Women Working So Much?" *Journal of Demographic Economics* 81 (1): 75–114.

Jones, Larry E., and Michele Tertilt. 2008. "An Economic History of Fertility in the U.S.: 1826–1960." In *Frontiers of Family Economics*, vol. 1, edited by Peter Rupert. Bingley, UK: Emerald Group Publishing, 165–230.

Kent, David A. 1995. "'Gone for a Soldier': Family Breakdown and the Demography of Desertion in a London Parish, 1750–91." *Local Population Studies* 45 (Autumn): 27–42.

Kinsey, Alfred C., Wardell B. Pomeroy, Clyde E. Martin, and Paul H. Gebhard. 1953. *Sexual Behavior in the Human Female*. Philadelphia: W. B. Saunders Company.

Kopecky, Karen A. 2011. "The Trend in Retirement." *International Economic Review* 52 (2): 287–316.

Kopp, Marie E. 1934. *Birth Control in Practice: Analysis of Ten Thousand Case Histories of the Birth Control Clinical Research Bureau*. New York: Robert M. Bride and Company.

Kremer, Michael. 1996. "Integrating Behavioral Choice into Epidemiological Models of AIDS." *Quarterly Journal of Economics* 111 (2): 549–573.

Malthus, Thomas R. 1798. *An Essay on the Principle of Population, as It Affects the Future Improvement of Society. With Remarks on the Speculations of Mr. Godwin, M. Condorcet, and Other Writers*. London: J. Johnson.

Manuelli, Rodolfo, and Ananth Seshadri. 2009. "Explaining International Fertility Differences." *Quarterly Journal of Economics* 124 (2): 771–807.

McGrattan, Ellen R., and Edward C. Prescott. 2017. "On Financing Retirement with an Aging Population." *Quantitative Economics* 8 (1): 75–115.

McLanahan, Sara, and Gary Sandefur. 1994. *Growing Up with a Single Parent: What Hurts, What Helps*. Cambridge, MA: Harvard University Press.

Mincer, Jacob. 1962. "Labor Force Participation of Married Women: A Study of Labor Supply." In *Aspects of Labor Economics: A Conference of the Universities–National Bureau Committee for Economic Research*, edited by Greg H. Lewis. Princeton, NJ: Princeton University Press, 63–105.

Mischel, Walter, Yuichi Shoda, and Monica L. Rodriguez. 1989. "Delay of Gratification in Children." *Science* 244 (4907): 933–938.

Mueller, Eva. 1976. "The Economic Value of Children in Peasant Agriculture." In *Population and Development: The Search for Selective Interventions*, edited by Ronald G. Ridker. Baltimore: Johns Hopkins University Press, 98–153.

Ogburn, William F. 1936. "Technology and Governmental Change." *Journal of Business of the University of Chicago* 9 (1): 1–13.

Ogburn, William F., and Meyer F. Nimkoff. 1955. *Technology and the Changing Family.* Boston: Houghton Mifflin.

Parente, Stephen L., Richard Rogerson, and Randall Wright. 2000. "Homework in Development Economics: Household Production and the Wealth of Nations." *Journal of Political Economy* 108 (4): 680–687.

Pareto, Vilfredo. 1897. *Cours d'économie politique.* Lausanne: F. Rouge.

Prescott, Edward C. 2004. "Why Do Americans Work So Much More than Europeans?" *Quarterly Review* 28 (1): 2–13.

Prescott, Edward C., and Graham V. Candler. 2008. "Calibration." In *The New Palgrave: A Dictionary of Economics*, 2nd ed., edited by Steven N. Durlauf and Lawrence E. Blume. London: Palgrave Macmillan, online.

Razin, Assaf, and Uri Ben-Zion. 1975. "An Intergenerational Model of Population Growth." *American Economic Review* 65 (5): 923–933.

Regalia, Ferdinando, and Jose-Victor Rios-Rull. 2001. "What Accounts for the Increase in the Number of Single Households?" Unpublished paper, Department of Economics, University of Minnesota.

Reid, Margaret G. 1934. *Economics of Household Production.* New York: John Wiley & Sons.

Rios-Rull, Jose-Victor. 1993. "Working in the Market, Working at Home, and the Acquisition of Skills: A General-Equilibrium Approach." *American Economic Review* 83 (4): 893–907.

Santos, Cezar, and David Weiss. 2016. "Why Not Settle Down Already? A Quantitative Analysis of the Delay in Marriage." *International Economic Review* 57 (2): 425–452.

Stone, Lawrence. 1977. *The Family, Sex and Marriage in England 1500–1800.* London: Weidenfeld and Nicolson.

Stone, Lawrence. 1993. *Broken Lives: Separation and Divorce in England, 1660–1857.* New York: Oxford University Press.

Suen, Richard M. H. 2006. "Technological Advance and the Growth in Health Care." Unpublished paper, Department of Economics, University of Leicester.

Vandenbroucke, Guillaume. 2008. "The American Frontier: Technology versus Immigration." *Review of Economic Dynamics* 11 (2): 283–301.

Vandenbroucke, Guillaume. 2014. "Fertility and Wars: The Case of World War I in France." *American Economic Journal: Macroeconomics* 6 (2): 108–136.

Varian, Hal R. 1978. *Microeconomic Analysis*. New York: Norton & Company.

Voena, Alessandra. 2015. "Yours, Mine and Ours: Do Divorce Laws Affect the Intertemporal Behavior of Married Couples?" *American Economic Review* 105 (8): 2295–2332.

Wilkinson, Richard G. 1973. *Poverty and Progress: Ecological Model of Economic Development*. London: Methuen.

Wrigley, Edward A., R. S. Davies, James E. Oeppen, and Roger S. Schofield. 1997. *English Population History from Family Reconstruction 1580–1837*. Cambridge, UK: Cambridge University Press.

Index

Abma, Joyce C., 220
Abortion, 200–201
Adverse selection problem, 248,
    249–251
Africa, AIDS in, 251
Age-specific divorce rate, 132
Age-specific fertility rate, 82
Age-specific marriage rate, 132
Aguiar, Mark, 2–3, 5, 52, 74,
    267
AIDS. *See also* HIV
    condom use and, 202
    decision to have unprotected sex,
        252–253
    effects of public policy on epidemic,
        254–255
    model set-up, 251–252
    presence of HIV in stationary
        equilibrium, 253–254
Aiyagari, S. Rao, 179
Albanesi, Stefania, 75–76, 126
Appliances. *See* Labor-saving household
    technologies
Asia, demographic transition in, 81
Assortative mating, 156–162. *See also*
    Positive assortative mating
    in Beckerian theory of marriage,
        176–177
    Gale-Shapley matching algorithm,
        157–160
    measuring strength of, 162

randomness in matching and,
    160–161

Baby boom, 83
    cause of, 126
    labor-saving household technologies
        and, 79, 94–95, 99–100, 102–103,
        126
    maternal medicine and, 103–108, 126
    mystery of, 86–89
    in OECD countries, 87, 100–103
Baby bust. *See* Fertility, baby bust
    (fertility decline, 1800–1990)
Barro, Robert J., 122–123, 124
Becker, Gary S.
    on child's rate of time preference, 232
    on defining leisure, 4
    economic analysis of discrimination
        and, 230
    household production theory and,
        37, 74
    model of fertility, 122–123, 124
    theory of marriage, 170–177, 178
Beckerian theory of marriage, 170–177
    assortative mating, 176–177
    stable and efficient matching,
        171–176
Benhabib, Jess, 75
Bentham, Jeremy, 199
Ben-Zion, Uri, 122, 123
Bethencourt, Carlos, 178

Bigamy, 151
Birth Control Clinical Research Bureau,
    202–203
Birth control movement, modern,
    200–201
Birthrate, crude, 80–81, 82
Bisin, Alberto, 232
Black, Dan A., 74
Black Death (plague), effect on
    population, 121
Blau, Francine D., 72, 73, 126–127, 179,
    230–231
Bliss shock, marriage and, 168–169
Blue-collar jobs, 65
drop relative to white-collar jobs, 269
    rise in married female labor-force
    participation and, 65–68
Boldrin, Michele, 125
Bosworth, Alfred W., 105
Braun, R. Anton, 266
Breastfeeding, fertility and, 105–106
Breast pump, 105–106
Bridgman, Benjamin, 74
Bronson, Mary Ann, 126
Browning, Martin, 178, 179
Budget line, in unisex single model of
    labor supply, 6–7, 9–12, 19, 24, 28,
    29, 30

Cain, Glenn, 72
Calibration
    advances in maternal medicine and
        female labor supply, 76–77
    decline in men's market hours, 76
    frequency of sex, data and model,
        222–223
    labor supply models, 35–37
    policy analysis, with calibrated
        economic models, 35, 36–37
    premarital sex model, 232
    unisex model of fertility, 127–128
Cancer, improvements in treatment and
    survival rates of, 236, 239

Candler, Graham V., 76
Capitalism/Protestantism, spirit of,
    223–230
    educational decisions and, 226–227
    endogenous response to
        technological progress and, 223
    industrialization and, 229–230
    instilling patience (self-control) in
        children and, 224–228
    investment in human and physical
        capital and, 223–224
    steady state, transitional dynamics
        and, 228–229
Caucutt, Elizabeth, 126, 179
Cavalcanti, Tiago V. de V., 75, 231, 232
Chiang, Alpha C., 275
Chiappori, Pierre-Andre, 178, 179
Childbearing, delayed, 126, 273
Child mortality, drop in, 125
Children. See also Fertility; Socialization
    of children
    being raised by a single mother,
        162–170
    cost of raising, 91–93, 95, 164, 270
    delayed childbearing, 126, 273
    premarital sex and socialization of,
        208–220
    single mothers and investment in,
        167, 168
Child well-being by family structure, 164
Childwite, 214
Churches. See Religion
Civil Rights Act, Titles VII and IX,
    184–185
Clandestine marriages, 151
Cobb-Douglas production function,
    37–38, 56, 57–58
Cobb-Douglas utility function, 14n3
Coen-Pirani, Daniele, 75
Cohabitation, 131, 179
Cohen, Ruth Schwartz, 2
Completed fertility rate, 82, 83–89
Comstock, Anthony, 200

Concavity and second-order condition
    for maximum, 276–277
Condoms
    for contraception, 198–199, 200, 201,
        202, 203
    decision to use for HIV prevention,
        252–253
    HIV infection prevention and, 251
Constant-elasticity-of-substitution (CES)
    production function, 55–56, 57–58
Constant relative risk aversion (crra)
    utility function, 13, 15
Constrained maximization, 278–281
Consumption
    Beckerian theory of marriage and,
        171–176
    consumption-fertility decision,
        89–100, 109–112, 118–122, 270
    consumption-leisure decision (for
        single), diagrammatical, 5–12
    consumption-leisure decision (for
        single), mathematical, 12–24
    consumption-leisure decision,
        married household, 25–35
    consumption possibilities frontier, 95
    economies of scale (for married life),
        129–130, 140–141, 147, 270–271
    of health care to extend life, 236–243
    of health insurance to extend life,
        247–251
    of home goods, 37–44
    of leisure goods in retirement,
        255–263
    of new drugs to extend life, 243–247
Contraception. See also Premarital sex
    effectiveness and use of, 201–204
    frequency of sex and, 221–223
    history of, 198–201
    improvements in, xii, 182–183, 231,
        272
    social change and technological
        progress in, 198–204
Corner solutions

instilling patience in children and,
    227–228
married female labor supply along the
    intensive margin and, 29–30
to maximization problems, 277–278
single household's labor supply
    (marriage model) and, 141–145
Correlation coefficients, 285–286
    Kendall rank, 285, 286
    Pearson, 285–286, 287–288
Cortes, Guido Matias, 73
Cotton, John, 183–184
Crude birthrate, 80–81, 82
Crude divorce rate, 132, 133
Crude marriage rate, 131–132, 133
Culture, effect of technological progress
    on, xii, 181-184, 271-272. See also
    Social change
Cumulative distribution functions
    (cdf's), 32–33, 282–283

Darwin, Charles, 117–118
DeGroot, Morris H., 275
Demographic transitions, 80–81,
    123–124, 127
de Nardi, Mariacristina, 125
Derivatives
    total derivative, 281–282
    total differentials, 281–282
Diaphragm, 199, 200
Difference equation. See Linear
    first-order difference equation
Differentials, total, 281–282
Dinkelman, Taryn, 75
Discount factor. See Patience; Time
    preference
Discrimination against women in the
    workplace, xiv, 47, 181–182,
    230–231, 271
    evolution of women's rights, 184–186
    gender wage gap and, 64–65, 113
    model of the evolution of women's
        rights, 186–191

Distribution functions, 282–286
  exponential distribution, 284
  Pareto distribution, 283–284
  uniform distribution, 284–285
Division of labor in household, 69–72
Divorce, 151–156
  cost of, 154
  decision to, 152–154
  in England (1660–1857), 151–152
  future trends in, 273–274
  no-fault, 154, 185–186
  remarriage and, 151, 152, 153,
    154–156
  trends in, xii, 132–137
Divorce rates
  age-specific, 132
  crude, 132, 133
Doepke, Matthias, 124, 125, 231, 232
Douching, 199–200
Drugs, development of new, 243–247,
    272–273
  drug monopolist's problem, 244–245
Duernecker, Georg, 74

Easterlin, Richard A., 126
Eckstein, Zvi, 125
Economic rationale for marriage,
    130–131. See also Money
Economies of scale (for married life),
    129–130, 140–141, 147, 270–271
Edgeworth-Pareto
    complements/substitutes
  definition, 60–61
  in marital production (assortative
    mating), 176, 177
  rise in married female supply, 61, 269
Edison, Thomas Alva, 45
Education
  assortative mating and, 137–139
  decision about how much education
    to acquire, 226
  decision about whether to become
    educated or not, 226–227

decline in marriage and level of, 137
female labor-force participation and,
    74
gender wage gap and level of, 63–65
kids, job choice, and, 112–113
Einav, Liran, 266
Elasticity of substitution, 56–58
England. See also United Kingdom
  demographic transition in, 81
  divorce in (1660–1857), 151–152
  Black Death, 121–122
  population and real wages in, 123
Envelope theorem, 42
Equal Employment Opportunity
    Commission (EEOC), 184, 185
Equal Pay Act (1963), 184, 185
Ermisch, John F., 74, 123, 179, 232
Europe. See also France; Switzerland;
    United Kingdom
  demographic transition in, 81
  World War I and drop in fertility in,
    108–109
Exercise, value of, 267–268
Exponential distribution, 284
Exponential utility, 14, 15
Extensive margin, married female labor
    supply along the, 25, 31–35

Family Limitation (Sanger), 200
Farm kitchen, 52, 53
Ferber, Marianne A., 73, 126–127, 179,
    230–231
Fernandez, Raquel, 73, 178, 231, 232
Fernandez-Villaverde, Jesus, 231
Fertility, xii, 79–128, 270. See also
    Unisex model of fertility
  baby boom and, 79, 86–89, 94–95
  baby boom in OECD countries and,
    100–103
  baby bust (fertility decline,
    1800–1990) and, 83–86, 94–95
  choice between jobs and kids and,
    112–117

definitions of fertility, 80–82
fertility and wars, 79–80, 108–112
increase in wages and, 79, 93–94,
 96–99
literature review on, 122–127
Malthus and, 117–122
model of fertility, 89–100
obstetric and pediatric medicine
 advances and, 104–108
old-age support and, 125–126
population dynamics, 100
Fertility rates
age-specific, 82
completed, 82, 83–89
crude, 80–81
general, 81–82
total, 82
Finkelstein, Amy, 266
First Industrial Revolution, xi. *See also*
 Second Industrial Revolution
end of Malthusian era and, 122
shift in attitudes and values and, 223,
 232
First-order condition, 276
Fogli, Alessandra, 73, 231
France
age pyramid in 1950, 110
baby boom in, 89, 91
fertility and World War I in, 108–112
nonmarital births in 19th-century,
 231–232
Paris in 1860s and nonmarital births,
 205
Frederick, Christine, 51, 52, 54
Fruits of Philosophy (Knowlton),
 199–200
Fuchs, Rachel G., 231
Functions, maximizing, 275–281.

Gale, David, 178
Gale-Shapley matching algorithm, 130,
 157–161
Galor, Oded, 73, 123–124

Gayle, George-Levi, 179
GDP, health-care spending as percent
 of, 236, 238
GDP per capita
cleaning/cooking and, 55, 56
market hours per week and, 5, 6
market hours per week for people age
 55 or older and, 257, 258
Gemici, Ahu, 179
Gender wage gap, xiv, 2, 62–65
brain and brawn, 65–68, 73
defined, 26
discrimination against women and,
 64–65, 113
education and, 63–65
jobs and kids, 112–117, 270
literature on, 72–73
measures of, 63–64
in OECD countries, 62–63
reduction of, 269
rise in married female labor-force
 participation (labor supply) and,
 46–47, 68
work experience and, 63–65
General fertility rate, 81–82
General Social Survey, 191
Gershoni, Naomi, 126
Gershuny, Jonathan, 74
Gilbreth, Frank B., 51–52
Gilead Sciences, 243
Godbeer, Richard, 231
Godey's Lady's Book, 131
Golan, Limor, 179
Goldin, Claudia, 73, 230
Government (and churches), desire to
 curtail premarital sex, 183, 213–220
Great Depression, baby boom and,
 86–89
Greenwood, Jeremy, 75, 124, 125, 126,
 178, 179, 231, 266
Griswold, Estelle, 200
Griswold v. Connecticut, 186, 200
Gronau, Reuben, 74

Guner, Nezih, 125, 126, 178, 179, 231
Gupta, Atul, 266

Hansen, Gary D., 124–125, 232
Happiness, success and, 232–233
Harms, Teresa Attracta, 74
Harvoni (drug for Hepatitis C), 243
Hayden, Mary, 215
Hayland, Elizabeth, 215
Hazan, Moshe, 231
Health, improvements in, xiii, 235,
　　236–243, 272–273
　AIDS, 235, 251–255, 266
　development of new drugs, 243–247
　health-care decision, 241
　health-care spending as percent of
　　GDP (income), 236, 238, 242
　health insurance, 247–251
　literature review on, 266
　model of life expectancy and
　　health-care spending, 237–241
Health insurance, 247–251
　adverse selection and, 249–251
　decision to purchase, 248–249
　market price for, 249
　single-payer mandate, 250–251
Heisig, Jan Paul, 75
Hepatitis C treatment (Harvoni), 243
Hercowitz, Zvi, 75
Herrendorf, Berthold, 74
Heston, Alan, 101
High-paying jobs, fertility and,
　　112–117
High School and Beyond Study (HSB),
　　163
Himes, Norman E., 202
HIV, 235. See also AIDS
　in Malawi, 251, 266
Home economics movement, 50–54
Home sector productivity, fertility and,
　　89–95
Household, impact of technological
　　progress on, xi-xiii, 270–274.

　　See also Labor-saving household
　　technologies
Household division of labor, 69–72
Household inputs
　divorce and decline in price of,
　　153–154
　marriage and decline in price of,
　　150–151
　married female labor supply and,
　　44–45
Household management, application of
　　scientific management to, 50–54
Household production theory, 1, 37–45,
　　73–74, 75
　housework decision, stage 1, 39–41
　literature on, 73–74, 75
　model set-up, 37–39
　rise in married female labor supply
　　and, 44–45, 55–62
　technology adoption decision, stage
　　2, 41–44
Household technology, choice of,
　　42–44. See also Labor-saving
　　household technologies
Housework. See also Nonmarket work
　decline in time spent on, 45-46,
　　50–51, 52, 55, 56
　determination of in household
　　production theory, 39–41, 58–59
　future decline in time spent on, 273
　Mrs. Verett, 49–50
　solutions for in marriage model, 144,
　　146
Human capital
　capitalism and, 223–224, 226–227
　single mothers and investment in,
　　163–164, 167, 168, 170
Hurst, Erik, 2–3, 5, 52, 74

Illegitimacy rate, 196–197
Illegitimacy ratio, 196, 197
Illegitimate births. See Nonmarital
　　births

Income. *See also* Income effect; Wages
  cross-sectional relationship between
    fertility and, 83–86
  impact of wage change in standard
    consumption-leisure diagram, 9–12
  size of baby boom and, 100–102
  Slutsky compensated shift in, 9–12,
    20, 24
  spending on children as fraction of
    household, 163-164
Income effect, 11–12
  adoption of household technology
    and, 44–45
  housework decision and, 40–41
  unisex model of labor supply with
    Leontief utility and, 22–24
  unisex model of labor supply with
    logarithmic utility, 14–20
  unisex model of labor supply with
    zero-income effect utility, 20–22
  Leontief utility function and, 24
  logarithmic utility function and,
    14–22
  in standard consumption-leisure
    diagram (properties), 8–9
  in unisex model of fertility, 96
  zero-income effect utility function
    and, 23, 24
Indirect utility function, 39
  in Gale-Shapley matching algorithm,
    157–158
  married household's, 146
  single household's, 141, 144–145
  technology adoption decision and,
    41–44
Industrialization, spirit of capitalism
    and, 229–230
Infant deaths, fertility and decline in,
    104
Infant formula, fertility and, 105
Information Age, xi
Intensive margin, married female labor
    supply along the, 25–30, 36

Isoelastic utility, 13, 15
Iyigun, Murat F., 179

Jaimovich, Nir, 73
Jones, Larry E., 73, 84, 85, 125

Kahn, Lawrence M., 73
Kendall rank correlation coefficient,
    285, 286
Kendall's τ, 101–102
Kent, David A., 151n5
Kinsey, Alfred C., 220
Knowles, John, 126
Knowlton, Charles, 199–200
Kocharkov, Georgi, 178, 232
Kopecky, Karen A., 266, 267
Kopp, Marie E., 202–203
Koreshkova, Tatyana, 266
Kremer, Michael, 266

Labor-saving household technologies
  baby boom and, xii, 79, 94–95,
    99–100, 102–103, 270
  decision to adopt and, 41–44
  decision to divorce and, 154
  diffusion through American
    households, 47–48, 50
  marriage decision and, 150
  married female labor-force
    participation and, 44–45, 59–62, 75
  quality-adjusted time prices for, 47,
    49, 50
  social change and, 182, 191
  washing machines, 47–50, 51
Labor supply, unisex single model of,
    5–24. *See also* Married female
    labor-force participation
  calibrating, 35–37
  with Leontief utility—no substitution
    effect, 22–24
  with logarithmic utility—equal income
    and substitution effects, 14–20
  mathematical formulations, 12–24

Labor supply (cont.)
    standard consumption-leisure
        diagram, 5–12
    with zero-income effect utility,
        20–22
Ladies' Home Journal (magazine),
    50, 51
Lagrange multiplier
    in Beckerian theory of marriage,
        174–175
    constrained maximization and,
        278–281
    retirement decision and,
        259–260
    as shadow value, 280–281
Lagrangian maximization problems,
    174–175, 259–261, 278–281
Land, Malthusian population
    model and, 80, 118, 119–122,
    124–125
Latin America, demographic transition
    in, 81
Laufer, Steven, 179
Laundry, time spent on before/after
    advent of washing machines,
    47–50, 51
Leibniz's rule, 282
Leisure
    defining, 4
    future rise in value of, 274
    hours spent by men and women on,
        2, 4–5
    standard consumption-leisure
        diagram, 5–12
Leisure goods, retirement, and increased
    consumption of, 236, 255–257,
    258–263, 267, 273, 274
Leon, Alexis, 75
Leontief production function, 56, 58
Leontief utility, unisex single model of
    labor supply with, 22–24
Leyrwite, 214
Life expectancy. *See* Longevity, increased

Linear first-order difference equation,
    286–289
    monotone convergence to the steady
        state, 289
    steady-state solution, 289
Linear-quadratic optimization
    problems, 111
Literature reviews
    on baby boom and baby bust,
        122–127
    on decline in marriage, 178–179
    on increased longevity and longer
        retirement, 266–267
    on rise in married female labor-force
        participation, 72–76
    on social change, 230–232
Logarithmic utility, 13, 15, 248n2
    unisex single model with logarithmic
        utility, 14–20
Longevity, increased, xiii, 235, 236, 237,
    272
    AIDS and, 251–255
    better health and, 236, 237, 238, 239
    development of new drugs and,
        243–247
    health insurance and, 247–251
    model of, 237–243
Love, as motive for marriage, 129, 130,
    148, 149–151. *See also*, Money, as
    motive
Lover, Samuel, 130
Low, Corinne, 126
Low-paying jobs, fertility and, 112–117
Lugauer, Steven, 75
Lynd, Robert and Helen, 52

Malawian HIV/AIDS epidemic, 251, 266
Malthus, Thomas R., 80, 117, 118, 121,
    122, 123
Malthusian equilibrium, unisex model
    of fertility and, 118–122
Malthusian population model, land
    and, 118, 119–122

Manuelli, Rodolfo E., 73, 123
Marginal rate of substitution, 8, 9
Marginal rate of transformation, 95
Market hours per week and GDP per
    capita, 5, 6
    for people age 55 or older, 257–258
Market work. *See also* Labor supply;
    Married female labor-force
    participation
    time spent in, 2–3, 5, 6, 257, 258
Marriage
    assortative mating and, 129, 130,
        137–139, 156–162, 176–177
    Beckerian theory of marriage,
        170–177
    decision to marry or not, 147–149
    decline in, xii, 129–151, 270–271
    definitions for marriage and divorce
        rates, 131–132
    divorce, 151–156
    economies of scale (in household
        maintenance), 129–130, 140–141,
        147, 270–271
    fraction of population that is married,
        149
    from economics to romance, 149–151
    indirect utility function for, 146
    literature review on, 178–179
    model of marriage, 140–151
    motives for marriage, 130–131
    remarriage, 154–156
    solutions for housework, market
        work, and consumption, 146
    trends in marriage and divorce,
        132–139
Marriage rates
    age-specific, 132
    crude, 131–132, 133
    race and, 179
Married female labor-force
    participation/labor supply, 1–77
    along the extensive margin, 25, 31–35
    along the intensive margin, 25–30, 36

attitudes toward, 182, 191–195, 231
brain and brawn, 65–68
calibrating labor supply models,
    35–37
gender wage gap and, 62–65, 68
history of, 45–55
household division of labor and,
    69–72
household production theory and,
    37–45, 55-62
increase in, xi, 25–35, 45–62, 68,
    269–270
jobs and kids, 112–117
literature review on, 72–76
maternal medicine and, 75, 76, 77
model of, 145–146
women's rights (model of) and, 186,
    189–191
Married households. *See* Marriage;
    Married female labor-force
    participation
Maternal death, decline in and fertility,
    103–104
Maternal medicine, baby boom and, xii,
    103–108, 126, 270
Maternal medicine, married female
    labor-force participation and, 75,
    76, 77
Maximizing a function, 275–281
    constrained maximization,
        278–280
    corner solutions, 277–278
    Lagrange multiplier as shadow value,
        280–281
    strict concavity and second-order
        condition for a maximum, 276–277
McCorvey, Norma, 200–201
McGrattan, Ellen R., 73, 267
McLanahan, Sara, 163, 179
Mean-preserving spread, 246, 247
Medela, 106
Median voter model of women's rights,
    xii, 181, 186–191, 271

Medicine
   baby boom and advances in obstetric
      and pediatric, 103–108
   development of new drugs, 235,
      243–247
   female labor-force participation and
      advances in maternal, 75, 76–77
   technological advances in, 272–273,
      274
Men
   hours spent on leisure, 2, 4–5
   hours spent working, 2–4, 76
Middletown (Indiana) study, daily
      housework in, 52, 55
Mincer, Jacob, 72
Mira, Pedro, 125
Mischel, Walter, 224
Money, as motive for marriage, 129,
      130–131, 147. *See also* Love, as
      motive
More Work for Mother (Cohen), 2
Mueller, Eva, 126
Mulligan, Casey B., 232

National Longitudinal Survey of Young
      Men and Women (NLSY), 163
National Survey of Families and
      Households (NSFH), 163
Needham, Orwell H., 105
Newenham, Joan, 215
Newport, George, 200
Nimkoff, Meyer F., 131, 133, 178
No-fault divorce, 154, 185–186
Nonmarital births
   as liability for church and state,
      213–220
   measures of, 196–198
   ∩-shaped pattern for teenage
      pregnancies over time, 183,
      207–208
   religious proscription against,
      183–184, 213–214
   sexual revolution and rise in, 272

   societal level of, 206–207
   in 19th-century France, 205, 231–232
Nonmarket work. *See also* Housework
   defined, 3
   time spent in, 2–3
North America, demographic transition
      in, 81

Obstetric medicine, fertility and
      advances in, 103–108
Occupations
   brain and brawn, 65–68, 73
   fertility and occupations chosen by
      women, 80, 112–117, 126, 270
   gender wage gap and, 63–65, 80, 117,
      126, 270
OECD countries
   baby boom in, 87, 100–103
   gender wage gap in, 62–63
Ogburn, William F., 131, 133, 178, 181,
      183, 195
Old-age pensions, fertility and, 125–126
Old-age social security
   fertility and, 125–126
   longer retirement and, 236, 273
   retirement decision and, 263–266
Olivetti, Claudia, 73, 75–76, 126, 231
Oreffice, Sonia, 178
Out-of-wedlock births. *See* Nonmarital
      births
The Overworked American (Schor), 2
Owen, Robert Dale, 199

Panel Study of Income Dynamics
      (PSID), 163
Parente, Stephen L., 75
Parents. *See* Single mothers;
      Socialization of children
Pareto, Vilfredo, 284
Pareto distribution, 283–284
Patience. *See also* Time preference
   instilling in children, 184, 227–228,
      232, 272

investment in human capital and,
223–224, 226–227
Stanford marshmallow test, 224
Pearson correlation coefficient,
285–286, 287–288
Pediatric medicine, fertility and
advances in, xii, 103–108, 126, 270
Peer-group effects, on premarital sex
decisions, 217–220
Penn World Table, 101
Pets, health-care spending on, 236, 240,
266, 268
Pill (contraceptive), 199, 200, 201, 203
Population dynamics, 100
Malthusian equilibrium and,
120–121
Positive assortative mating, 129, 130,
271. *See also* Assortative mating
Beckerian theory of marriage and,
170–171, 176–177
future of, 273–274
rise in, 137–139, 161–162
Poverty, single motherhood and
percentage of children living in,
163–164, 165
Poverty cycle, single motherhood and,
130, 165–170
Premarital sex, 182–184, 231, 272
calibrating model of, 232
church and state, 183–184, 213–220
model of, 204–208, 231
parents' socialization problem about,
212–213
peer-group effects and, 217–220
sexual revolution and, 195–196
socialization of children and,
183–184, 208–220, 272
societal level of, 206–207
technological progress in
contraception and, 198–204
teenage girls' decision-making
regarding, 205–206
Prenuptial pregnancy ratio, 197–198

Prescott, Edward C., 76, 124–125, 232,
267
Price
appliances (time), 47, 49
health insurance, 249
household inputs (female labor
supply), 37–45
household inputs (marriage model),
140–151, 154
household technology adoption and,
43–44
leisure goods, 257, 262
Private savings accounts, 266, 267
Probability density function, 283
Production function
Cobb-Douglas, 37–38, 56,
57–58
constant-elasticity-of-substitution
(CES), 55–58
for raising children, 91–92
*The Protestant Ethic and the Spirit of
Capitalism* (Weber), 223

Quadratic utility function, 14, 15,
109–110
Quintana-Domeque, Climent, 178

Race, marriage rates and, 179
Rauh, Christopher, 179
Razin, Assaf, 122, 123
Real per-capita GDP
average market hours and, 5, 6,
257, 258
cooking/cleaning and, 55, 56
Regalia, Ferdinando, 179
Reid, Margaret G., 37, 73–74
Religion
premarital sex and socialization of
children, 213–216, 218–220
proscription against premarital sex
and out-of-wedlock births,
183–184
Remarriage, 154–156

Retirement, longer, xiii, 236, 273, 274
  leisure goods and, 256–257
  literature review on, 267
  model of, 258–263
  old-age social security and,
    263–266
  private savings accounts and, 266
  retirement decision, 259–263
  trends in retirement, 255–263
Rios-Rull, Jose-Victor, 75, 178, 179
Roe v. Wade, 186, 200–201
Rogerson, Richard, 75

Same-sex couples, stay-at-home
    partners and, 69, 74–75, 131
Same-sex marriages, 131
Sandefur, Gary, 163, 179
Sanders, Seth G., 74
Sanger, Margaret, 200
Santos, Cezar, 178, 232
Schor, Juliet, 2
Scientific management, applied to
    household management, 50–54
Second Industrial Revolution, xi.
    See also First Industrial Revolution
  decline in housework/rise in married
    female labor supply and, 47–54
Second-order condition, 276–277
Self-control (patience), model of
    instilling, 224–230
Seshadri, Ananth, 75, 123, 124, 126
Sex. See also Premarital sex
  AIDS and decision to have
    unprotected, 252–253
Sex, frequency of, 220–223
  matching model with data
    (calibration), 222–223
Sexual harassment, women's rights and,
    185
Sexual mores, impact of advances in
    contraception on, xii, 182–184, 272
Sexual revolution, 182–183, 195–196,
    272

Shame, premarital sex and inculcation
    of, 183–184, 208–213, 219,
    231, 272
Shapley, Lloyd S., 178
Shultz v. Wheaton Glass, 184–185
Similac (baby formula), 105
Single households, in model of
    marriage, 141–145
  corner solution (market work),
    142–143
  decision to marry or not, 147–148
  interior solution (market work), 142
  single household's indirect utility
    function, 144
  solutions for housework, market
    work, and consumption, 144
Single mothers, 130, 162–170, 179
  background, 163–165
  being raised by single mother
    (model), 165–170
  investment in children's human
    capital and, 167
  poverty cycle and, 130, 163, 167, 170
  steady-state fraction of women who
    are single mothers, 169–170
Single-payer mandate, 250–251
Single women, number of sexual
    partners, 195, 197
Siu, Henry E., 73
Slutsky compensated shift in income,
    9–12, 19–20, 24
Social change, xii, 181–233, 271–272
  attitudes toward married women
    working, 182, 191–195, 231
  educational decisions (and dawning
    of capitalism), 226–227
  frequency of sex, 220–223
  instilling patience (self-control) in
    children, 184, 224–230
  literature review on, 230–232
  model of premarital sex, 182–183,
    204–208
  nonmarital births, 196–198

premarital sex and socialization of children, 183–184, 208–220, 231
sexual revolution (1900–2000), 182–183, 195–196
spirit of capitalism, 223–230, 232
technological progress in contraception, 182–183, 198–204, 231
women's rights in the workplace, 181–182, 184–191, 230–231, 271
Social Change with Respect to Culture and National Origin (Ogburn), 195
Socialization of children
instilling patience (self-control), 224–230
premarital sex and, xii, 183–184, 208–220, 272
Social Security Act (1935), 263, 273
Socioeconomic class, positive assortative mating and, 129, 137–139, 271
Sons, attitudes toward married women working and, 182, 191–195, 231
Soytas, Mehmet A., 179
Standard of living, technological progress and increase in, 235, 236
Stanford marshmallow test, 224
Stanton, Elizabeth Cady, xii, 45
Stone, Hannah, 200
Stone, Lawrence, 151n4, 215
Substitution effect, 11, 12
housework decision and, 40–41
unisex labor supply model with Leontief utility, 22–24
unisex labor supply model with just-substitution (or zero-income) effect utility, 20–22
unisex labor supply model with logarithmic utility, 14–20
Suen, Richard M. H., 266
Summers, Robert, 101
Switzerland, baby boom in, 89, 92, 101

Tavares, Jose, 75, 231
Tax policy analysis, 36–37, 180, 250–251
Taylor, Lowell J., 74
Taylor, Sidney, 185
Technological progress, xi-xiii, 269–274. *See also* Labor-saving household technologies; Medicine
baby boom, baby bust and, 79, 94–95, 99–100, 107–108, 126
brain, brawn, and industrialization, 65–68, 73
contraception/premarital sex and, 182–183, 201–204, 206–207, 219–220
decline in marriage and, 129, 149–151
housework, female labor-force participation and, 44–45, 47–55, 59–62, 68, 73, 74, 75–76
increase in longevity and, 235, 242–243
industrialization and the spirit of capitalism, 223–224, 229–230, 232
longer retirement and, 255, 262
rise in positive assortative mating and, 161–162
women's rights and, 181–182, 191
Tertilt, Michele, 84, 85, 231
Threshold values (work/don't work), 32, 33
Time, measuring allocation of, 2–5
Time preference, 223–224, 232. *See also* Patience
determination of, 227–228
transitional dynamics toward steady state, 228–229
Title IX of the Education Amendments (1972), 185
Title VII of the Civil Rights Act of 1964, 184–185
Total derivative, 281–282
Total differentials, 281–282
Total fertility rate, 82

Uniform distribution, 284–285
Unisex model of fertility, 89–100.
   See also Fertility
   advances in obstetric and pediatric
      medicines and, 106–108
   baby boom and baby bust, 94–95
   calibrating, 127–128
   choosing between jobs and kids,
      113–117
   consumption possibilities frontier, 95
   determination of fertility, 96–100
   diagrammatic transliteration of,
      95–100
   indifference curves over consumption
      and fertility, 96
   Malthusian equilibrium and,
      118–122
   population dynamics, 100
   wars and, 109–112
Unisex single model of labor supply.
   See Labor supply, unisex single
      model of
United Kingdom, wars and fertility in,
   88–89, 90. *See also* England
United States
   advances in maternal health, 103–106
   baby boom in, 83n1, 86–89
   child care time by mothers in, 164,
      166
   child well-being by family structure,
      163–165
   diffusion of basic facilities and
      appliances, 47, 48
   spending on children as fraction of
      household income, 164
   technological progress in
      contraception, 199–204
   time price of appliances, 47, 49
   trends in female labor-force
      participation, 45–46, 65, 66
   trends in fertility, 83–86, 87, 124
   trends in health, 236, 237, 238, 239,
      240

   trends in premarital sex/nonmarital
      births, 195–197
   trends in retirement, 255–257
   trends in time use, 2–5
   women's rights, milestones, 184–186
U.S. v. One Package, 185, 186, 200
Utility function and assumptions
   imposed on, 12–14

Van Baer, Karl Ernst, 199
Vandenbroucke, Guillaume, 75, 108,
   124, 125, 126
Verdier, Thierry, 232
Vinson, Mechelle, 185
Voena, Alessandra, 178

Wages. *See also* Gender wage gap;
   Income
   decision to divorce and growth in,
      154
   fertility and increase in, 79, 84,
      93–95, 97–99
   impact of change in on standard
      consumption-leisure diagram, 9–12
   impact of change in on unisex labor
      supply model with Leontief utility,
      24
   impact of change in on unisex labor
      supply model with logarithmic
      utility, 18–20
   impact with zero-income effect utility
      (wage elasticity of labor supply), 22
   increased longevity and, 237, 242
   kids, jobs and, 116–117
   Malthus and, 121–122
   marriage decision and, 149–150
   married female labor-force
      participation and, 27–28, 33–35
   new drug development and rising,
      246–247
   retirement and, 236, 262–263
   rise in married female labor supply
      and rise in general level of, 45–46

rise in positive assortative mating and
    rise in, 161–162
wage penalties for woman taking time
    off, 112–113
Wars, effect on fertility, 79–80,
    108–112
Washing machines, 47–50, 51
Weber, Max, 223
Weil, David N., 73, 123–124
Weiss, David, 178, 231
Weiss, Yoram, 178, 179
Wheaton Glass Company, 184–185
White-collar jobs, 65
    rise in married female labor-force
        participation and, 65–68
    rise relative to blue-collar jobs, 269
Wilkinson, Richard G., 125
Winkler, Anne E., 72, 73, 126-127, 179,
    230–231
Wolpin, Kenneth I., 125
Women. See also Gender wage gap;
    Married female; Single mothers
    attitudes toward working, 73, 182,
        191–195, 231
    discrimination against, xiv, 47, 63,
        65, 230–231
    hours spent on leisure, 2, 4–5
    hours spent working, 2–4
    living arrangements for, 135–136
    work and delayed childbearing, 126,
        273
Women's Bureau, 184, 185
Women's rights in the workplace, xii,
    184–191, 271
    decision to work or not, 187–189
    milestones in, 184–185
    model of, 186–191
    vote to grant rights or not, 189–191
Wong, Joyce Chen, 178
Work experience, gender wage gap and
    differences in, 63–65
Working, hours spent, 2–4, 5
World War I, fertility and, 108–112

World War II
    baby boom and, 86–88
    trends in marriage and divorce since,
        132–134
Wright, Randall, 75
Wrigley, Edward A., 198

Yorukoglu, Mehmet, 75

Zero-income effect utility function,
    unisex single model of labor supply,
    20–22
Zilibotti, Fabrizio, 232
Zoabi, Hosny, 231